FUNGI IN THE ANCIENT WORLD

How Mushrooms, Mildews, Molds, and Yeast
Shaped the Early Civilizations of Europe,
the Mediterranean, and the Near East

Frank Matthews Dugan

U.S. Department of Agriculture, Agricultural Research Service
Washington State University, Pullman

APS
PRESS

The American Phytopathological Society
St. Paul, Minnesota

Cover: front, Greek votive or funeral stele, circa 470–460 B.C., Louvre (Réunion des
Musées Nationaux/Art Resource, NY); and (left to right), *Amanita caesarea* (Dufour
1891); *Piptoporus betulinus* (Dufour 1891); *Amanita muscaria* (Dufour 1891); back (left
to right), *Phellinus igniarius* (Dufour 1891); *Amanita caesarea* (courtesy VincEnzo
Ricceri); *Fomes fomentarius* (Cordier 1870).

Library of Congress Control Number: 2007941859
International Standard Book Number: 978-0-89054-361-0

Published 2008 by The American Phytopathological Society

Printed in the United States of America on acid-free paper

The American Phytopathological Society
3340 Pilot Knob Road
St. Paul, Minnesota 55121, U.S.A.

Preface

This book surveys the diverse roles of fungi in ancient societies, from the beginnings of the Neolithic until the advent of medieval times. Cultures of the ancient Near East (including the Fertile Crescent, Asia Minor, and Egypt), the Mediterranean, and Europe are the primary focus, but some minor excursions are made farther afield temporally and geographically.

Mycologists and plant pathologists, accustomed to the rigors of hypothesis testing and the use of controls, will find here an abundance of literature produced in conformity with such principles. In addition, numerous publications regarding the roles of fungi in myth, folklore, and ritual have been written by historians, mythographers, classicists, and linguists, whose methods justifiably differ from those of the biological sciences. I have attempted to include a wide assortment of opinion on especially controversial topics, such as the extent to which fungi functioned as ritual "entheogens" and whether mycotoxins were responsible for certain ancient historical events. Fermented beverages are shown as integral to the politics and social structures of the most ancient monarchies. The supremely important coevolution of fungi with agriculture is placed in context with archaeobotany, plant landraces and other living ancient crops, and fossil fungi and living fungal germplasm. These and other aspects are presented to mycologists, plant pathologists, historians, and others who might gain profit and pleasure from a "spore's eye view" of ancient history!

Acknowledgments

I am much indebted to Dean Glawe (Department of Plant Pathology, Washington State University, and College of Forest Resources, University of Washington) and Tim Paulitz (USDA-Agricultural Research Service and Department of Plant Pathology, Washington State University) for advice on mycological aspects of this study, to Carol Thomas (Department of History, University of Washington) for constructive review of the historical context and ancient sources, and to Guntis Šmidchens (Department of Scandinavian Studies, University of Washington) for review of sections on folklore. Owen Ewald (Classics Department, Seattle Pacific University) provided guidance on relevant passages in Cassius Dio. Research Leader Richard Hannan (USDA-Agricultural Research Service, Washington State University) provided support and encouragement, especially in matters pertaining to plant germplasm and the history of beer and mead. The staff at APS PRESS was always helpful. I take sole credit for errors and omissions.

List of Figures

Table of Contents

Introduction

The history of interactions between humans and fungi is a fascinating story, integral to the histories of agriculture, medicine, the culinary arts, and even religion. Because fungi produce few structures equivalent to bone, shell, arthropod exoskeletons, or the more durable portions of higher plants, there is less direct archaeological evidence for fungi than for some other groups of organisms. However, modern research has been able to detect fungi in a multitude of archaeological contexts, and much can be plausibly inferred even in the absence of fungal structures. Ancient peoples themselves often recorded pertinent observations on parchment, papyrus, cuneiform tablets, pottery, or other media. Romans or Greeks, for example, referred to *boleti* and *suilli* (edible mushrooms), *erysibe* (mildew on plants), *agaricum* (medicinal fungus), *mucor* (mold), *fermentum* (yeast), and *fungi* (mushrooms and toadstools).

By assembling evidence from multiple sources, we can clearly discern impacts of fungi throughout ancient times. Human institutions and culture were dramatically altered in multifaceted ways by the fungal kingdom. In some instances, such as enology, baking, and ritual feasting, institutions were entirely predicated on the existence of those fungi called yeasts. In other instances, such as the production of mycotoxins in food stores; crop failure from rusts, mildews, or other plant diseases; and the confusion between fungal ringworm and true leprosy, both individuals and entire settlements probably suffered irreparable negative impacts.

Most impacts of fungi on human societies are closely connected with plants, agriculture, and associated activities. An understanding of the main elements of the Neolithic is essential to thinking constructively about fungi and early societies. The Upper Paleolithic witnessed an increase in human populations and more sophisticated technology, such as grinding tools for processing plant foods (Bar-Yosef 2002). Although remains of food plants themselves are not well represented in Paleolithic sites, they are well attested in archaeological contexts from the Neolithic starting from about 10,000 B.C. Intense utilization of higher plants as food, especially those high in carbohydrates, started in the Neolithic, as indicated by increased use of stone grinding tools. Also at this time, teeth from human remains start bearing the signs of dental caries (linked to diets high in carbohydrates) plus wear from poorly ground flour (Richards 2002).

The Neolithic era began in the Fertile Crescent, the great arc comprising the eastern Mediterranean, portions of what is now southeastern Turkey, and the lands between and adjacent to the Tigris and Euphrates rivers. The Neolithic spread outward to Egypt, Iran, western Anatolia, and the coasts and islands of the Aegean. Somewhat after 8000 B.C., the "foundation crops" associated with the ancient Near East were all well established in Mesopotamia and adjacent regions (Bellwood 2005, Zohary and Hopf 2000). Sowing of these plants (emmer, einkorn, and eventually bread wheat, plus barley, pea, chickpea, lentil, and others) and their storage as food or seed provided opportunities for various plant-pathogenic fungi and storage fungi to adapt to agricultural systems. The degree to which farming spread by the migration of peoples versus by the diffusion of agricultural technology has been much debated (e.g., Armelagos and Harper 2005a,b; Bellwood 2005; Brown 1999; Pinhasi and Pluciennik 2004; Richards 2003), but it is generally agreed that advanced agricultural systems first dominated in the ancient Near East and spread northward and westward into Europe. Around 7000 B.C., farmers settled in the alluvial basins of Greece (Perles 2001). Well before 5000 B.C., emmer and einkorn were grown in middle Europe. Agriculture was established in western Europe by about 4000 B.C. and after several hundred more years, in the more northerly Baltic and British Isles (Bellwood 2005). By the time farming was established in northern Europe, much of the ancient Near East already had undergone profound changes, including urban collapse and renewal, cycles of deterioration and restoration of agricultural lands, and shifts in agricultural economies (Fall et al. 1998).

This quick sketch indicates, because of its primarily agricultural focus, a context appropriate to the biases and preoccupations of a professional plant pathologist. I especially concentrate on information pertinent to plant-pathogenic fungi and other fungi affecting human diet and health. My own career with extensive research collections of higher plants and fungi has no doubt skewed the presentation toward fungus-plant interactions. To this day, persons seeking germplasm for creation of plant cultivars resistant to disease explore the Near East, the Balkans, and other geographic areas included in this study (see Lenné and Wood 1991, Leppik 1970). My colleagues at Washington State University often do research on plant diseases that were the objects of fear and propitiation by the ancients. In this volume, however, in addition to agriculture, I have included much that relates to folklore, religion, and ritual. Several topics are inherently controversial, and with these I have tried to describe as fairly as possible the range of opinion without entirely obscuring my own biases.

Some historians, and especially classicists, will no doubt find fault with the manner in which certain material is documented. Considerable reliance is placed on secondary sources, especially when the primary sources are in the languages of antiquity. It should be kept in mind that, with the occasional exception of some rusty Latin, most readers (including myself) lack familiarity with ancient languages. Moreover, the secondary sources frequently provide explanatory context quite lacking in the writings of the ancients themselves. In most instances, the sources cited provide sufficient information so that original material in Latin, Greek, Hebrew, cuneiform, and hieroglyphic scripts can be consulted. Readers lacking familiarity with Latin or Greek but wishing to delve further into some original materials are referred to the Loeb Classical Library (Harvard University Press), whose volumes, sometimes cited directly herein, contain both the full Greek or Latin text plus an English translation.

This review confines itself to evidence from Europe, the Mediterranean, and adjacent west Asia from the Neolithic through classical antiquity. This restricts our scope to the impact of fungi on early societies of the Western tradition and its immediate predecessors. For times of much greater antiquity (including those antecedent to the evolution of our species), the reader is referred to Taylor and Osborn (1996). Schultes (1998), Wasson (1980), and Morgan (1995) cover some interesting aspects of fungi and human societies in the Americas. For a glimpse of the culinary traditions of East Asia, see Huang (2000) and McGovern et al. (2004). The early history of fermented beverages in South America, Africa, and East Asia is summarized by Hornsey (2003). Orlob (1971, 1973) addresses aspects pertinent to plant pathology in the Far East, pre-Columbian Americas, and medieval times. Capsule histories of the spread of agriculture in geographic areas outside the Near East and Europe are provided by Bellwood (2005). Toporov (1985) reviews conceptions about mushrooms in myth and folklore worldwide, and Dugan (in press) summarizes fungi in folklore and fairy tales. Money (2007) gives a concise and entertaining overview of the historical impact of plant-pathogenic fungi worldwide. Persons in an extreme hurry may wish to consult Baker (1965), who compresses our subject and more into 35 small pages.

A knowledge of fungal identification and systematics is not essential for understanding what follows, but motivated readers can consult Dugan (2006), McLaughlin et al. (2001a,b), or for an exquisitely humorous introduction, Money (2002). The general reader can be content with knowing that mushrooms and toadstools, conks (those woody, hooflike or shelflike things growing on trees), molds, and yeasts (including the yeasts used for making bread, beer, and wine) are all fungi. Microscopic fungi often cause plant diseases (rusts, smuts, mildews, leaf spots, stem or branch cankers, root rots, etc.) or diseases of humans and animals (e.g., ringworm, yeast infections, allergies, or athlete's foot). Many microscopic fungi harmlessly decompose leaves and other litter.

Each chapter is divided into two sections: a brief narrative synopsis in which references are omitted, and a longer section providing documentation. These sections may be read sequentially or independently. It should be stressed at the outset that much of the subject matter is controversial, conclusions are often conjectural, and only the most credulous will accept the summary narratives without consulting at least some of the documentation.

Fungi in Baking and Brewing

The archaeological evidence for baking goes back at least 9,000 years. Documentation from cuneiform and hieroglyphic scripts supports the great antiquity of the use of leavening agents (yeasts). Egyptians, Mesopotamians, and early Europeans all evolved analogous practices in which byproducts of beverage fermentation were used as starters for the dough in leavened breads. Indeed, beverage fermentation may have preceded baking historically. By the time of urban cultures in Sumer and Egypt, fermentation of alcoholic beverages and baking of leavened bread were both well established, common practices often conducted on a large scale and with distinct regional traditions. Baking probably became easier and more widespread as bread wheat replaced emmer and einkorn wheat (both of which have seeds that retain their hulls at harvest, in contrast to the "naked" seeds of bread wheat). By the time the Greco-Roman civilizations were established as the dominant cultures of the Mediterranean, leavened bread was established throughout that region and beyond, a guild of bakers had been founded in Rome, and various alternatives for the production of leavening agents had been well described in Roman literature.

Brewing of beers or mead certainly coevolved with bread baking and may have preceded it. Wine production was more constrained by climate, and in northern Europe wine long remained a commodity obtained only via trade. Even in the rather porous ceramics of the time, wine kept better than beer and mead and became a staple of trade throughout the Mediterranean region. Fermented beverages were valuable commodities and generated a spectrum of cultural practices, mythologies, and rituals throughout Mesopotamia, Egypt, the Mediterranean, and Europe. There was even a trade in medicated wines, roughly analogous to the ethanol-opiate concoctions of nineteenth-century patent medicines. Writers both ancient and modern have been fascinated with the impact of alcohol on society, rituals performed in conjunction with fermented products, the deities evoked in such rituals, their connotations of fertility and sexuality, and even the echoes of ritual human sacrifice. Some modern writers have (even when presumably sober) promulgated the idea that production of fermented beverages was the prime incentive for the establishment of agriculture. Although contemporary observers frequently associate alcohol abuse with breakdown in social order, a well-supported hypothesis regards ritual drinking at feasts to be a primary mechanism for binding ancient warriors or other notables to their monarch or chief (see Some Additional Hypotheses Regarding the Impact of Fungi in Ancient Times).

Baking and brewing (and their biotic agents, primarily the yeast *Saccharomyces cerevisiae*) arguably represent not just the most ancient but the most important biotechnologies of all time. Controversies over details of ancient brewing methodologies have provided researchers with an incentive to duplicate ancient practices and (by intensive sampling no doubt) pass judgment on the taste and ethanol content of the results.

Documentation: Fungi in Baking and Brewing

Baking: Light Bread and Heavy Ovens

Leavened bread was probably present in the earliest complex societies of the ancient Near East. Some authors have suggested that bread was baked as early as 10,000 B.C. (Jenson 1998, citing Wood 1989), but most place baking, especially of leavened bread, sometime later in the Neolithic or early Bronze Age. Ancient leavening agents would have been sourdoughs containing both yeast and lactic acid bacteria (Jenson 1998, citing Wood 1989). The Egyptians were probably the first to experiment with leavened breads, and they kneaded dough with their feet in the larger bakeries (Brothwell and Brothwell 1969). The most comprehensive review of cooking in ancient Mesopotamia (Bottéro 2004) attests the early use of yeasts as leavening agents for bread. Leavened and unleavened bread coexisted in the Mesopotamian diet. The ancient hearth used for baking unleavened bread was well known to the Sumerians, whose word for it, *tinûru*, was probably borrowed and has philological equivalents in Turkish, Arabic, Iranian, and Indic. Leavened bread is specifically referenced in Mesopotamian epic poetry; it was prepared with a little beer or "left-over soured soup" (Bottéro 2004), and the oven used for baking it differed from the *tinûru*. The historical and contemporary distribution of ovens versus griddles in western Asia and northeastern Africa is well documented (Lyons and D'Andrea 2003), ovens being associated with cereals of high gluten content, especially wheat, and leavening. Domed ovens are found as early as 7000 B.C. in Turkey and Iraq, but there is no direct evidence of their being used for leavened bread (Lyons and D'Andrea 2003). The origins of baking or roasting of seeds of wild cereals and grasses in ovenlike stone-lined hearths can be dated well prior to the Neolithic (Piperno et al. 2004). Wood (2000) reviews ovens, including earthen ovens, in European (mostly Celtic) prehistory.

Mortimer (2000) is seemingly at variance with the above authors: "It appears that bread making dates back at least 6000 years, but use of leavening, which required the development of suitable cereal grains with easily removable hulls, gluten, and the introduction of yeast cells, did not appear until around 500 B.C." Mortimer cites McGee (1984), but McGee actually says that "raised bread . . . seems to have developed in Egypt around 4000 B.C. . . . Leavened bread seems to have been a rather late arrival along the northern rim of the Mediterranean. The new wheat [i.e., without hulls, or "naked" wheat] was not grown in Greece until about 400 B.C."

Egypt was likely the locale for the first leavened bread, but baking may have been developed only subsequent to primitive but intentional fermentation of grains (Hornsey 2003). Wood (2000), quoting Pliny, notes that beer foam was used for leaven in ancient Gaul and Spain and that the resultant bread was of "a lighter kind" than others, possibly on par with or superior to Roman bread. Wood emphasizes that the "interdependence between the grain and the yeast, between bread and fermenting" was established in the earliest times. McGee (1984, p. 280), also citing Pliny, relates the standard Roman method of obtaining yeast:

> The simplest and oldest was to save a piece of old dough. During the grape harvest, however, millet or wheat bran was mixed with grape juice, allowed to become contaminated by air, and then dried in the sun. The resulting cakes were soaked in water when needed. Barley cakes were sealed in clay pots until they became sour and yeasty, then dispersed in water.

Use of yeast became thoroughly established in Rome (against some resistance from conservative citizens, who thought light loaves unhealthy) when the bakers' guild, the *Collegium Pistorum,* was established in 168 B.C. A capsule history of this guild, plus a concise but comprehensive review of Roman bread and a more extended translation of Pliny's comments on the production of leavening agents, is available in Breatann (n.d.). The Greco-Roman period saw multiple innovations in oven design and in the shape and decoration of loaves (often, for ceremonial purposes, in the shapes of male or female genitalia), with social aspects of baking culminating in the activities and regulations of the castelike *Collegium Pistorum* (Toussaint-Samat 1992).

Brewing: Fermentation Products and Deities

Fermented beverages were present in the earliest complex societies of the ancient Near East and were symbolized in early writing systems

(Figure 1). Bottéro (2004) attests the early use of yeasts as fermenting agents for beer and wine. The ideogram for beer, present in Sumerian cuneiform from the end of the fourth millennium B.C., shows a large vessel containing water and grain, but most of the vocabulary describing beer production was not indigenous to the Sumerian language, and so it is plausible that the Sumerians learned the techniques from another people (Bottéro 2004). Beer production likely antedated complex societies in both Mesopotamia and Egypt (Jennings et al. 2005). Brewing vats of very early date (ca. 3500 B.C. and later) have been excavated in Pre-Dynastic and Early Dynastic Egypt (Darby et al. 1977, Geller 1992, Mills 1992). Beer was the national drink in ancient Egypt as well as a staple food item. In ancient Egyptian practice, and still in the modern brewing of *bouza*, the folk brew in contemporary Egypt, partially baked loaves were combined with sprouted grain (malt) to provide the substrate for fermentation. Figures from ancient tombs showing the making of bread and beer, and photographs, reconstructions, and excavation plans of brewery vats give an idea of the various scales of production (Darby et al. 1977, Geller 1992).

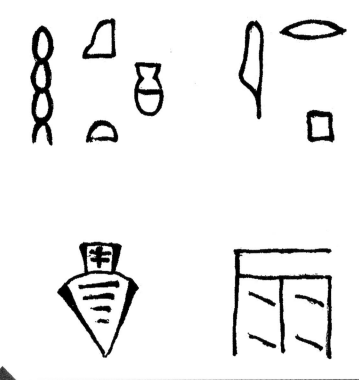

FIGURE 1 Fermented beverages as depicted in ancient writing systems. Top row: Egyptian hieroglyphics for beer (left) and wine (right). Bottom row: Mesopotamian cuneiform for beer (left) and Mycenaean Linear B for wine (right).

Several of the many terms for types of beer have been translated (sweet beer, dark beer, garnished beer, beer that does not sour, beer of the goddess Ma^cat, etc.), as have anecdotes from ancient writings involving beer taverns, drunken students, and overindulgence at feasts (Darby et al. 1977, Geller 1992). Herodotus—"father of history" (Cicero) or "father of lies" (Thucydides) (the ancients were not unanimous in their opinion of Herodotus)—noted the identification of the Greek god Dionysius with the Egyptian deity Osiris, but the Egyptians considered the goddess Hathor and her junior partner Menqet as the inventors of brewing (Geller 1992), and apparently women were prominent as brewers, just as they were in Babylonia (Saggs 2005).

Geller (1992) reviews the extensive role of beer in Egyptian mortuary practices. The dead apparently appreciated an ample supply of beverages, and the Egyptians and other cultures obliged them. A lavish burial from the Celtic Hallstaat culture contained a large cauldron with residues of mead (Freeman 2006). Wine was also placed in tombs. King Tutankhamen apparently opted for red over white (Guasch-Jane et al. 2004).

Beer remained the dominant beverage in Mesopotamia until at least the end of the first millennium B.C. (Bottéro 2004). It was brewed substantially as we brew it today, i.e., by allowing grain to sprout and allowing the mash to ferment. Hops were not used, but other flavorings could be added.[1] Partially fermented barley porridge was a form of "fast food" in the ancient Near East (Pollock 2003). Jennings et al. (2005) review the literature and summarize production processes in ancient Egypt. They state that although "yeast cells have been identified in residues, the specific varieties of yeast used remain unknown." There is evidence for lactic acid fermentation in some Egyptian beverages and also evidence that Egyptian and Mesopotamian practices for brewing differed in some details. The crumbling of bread into water has been widely accepted as an ancient practice for making beer in Egypt, but Williams (1996) reports evidence that malting and heating of cereal grains, followed by mixing with sprouted, unheated grains, preceded decanting and fermentation. By contrast, Ishida (2002), who used data from Egyptian folklore as well as archaeological evidence to reconstruct Egyptian brewing practices, emphasizes the use of beer bread as a starter culture. In experiments, Ishida brewed beer by analogous methods: "The taste of the beer was considered good" with "good body"—certainly a compliment, considering that the author was from the technology department of the Kirin Brewery! Scheer (2004) conducted similar experiments using bread as an ingredient to recreate a Babylonian beer, "Ninkasi brew." Bread wheat

---

[1] Hornsey (2003) and Wilson (1975) discuss the purported evidence for hops in brewing before medieval times. Both authors are skeptical of the premedieval use of hops in brewing but provide ample evidence for it in the medieval period.

is hypothesized to have replaced emmer wheat as the raw material for beverage production sometime between the Old and Middle Kingdoms (Ishida 2002).

Beer's cousin, mead (brewed with honey as the principal carbohydrate source), may well have preceded wine in Europe (Hornsey 2003). However, archaeological evidence for mead is ambiguous, because such evidence (remains of beeswax, certain kinds of pollen) may only imply that honey was added to sweeten another beverage (Hornsey 2003). For example, *corma* was a honey-flavored beer popular among the lower classes in Gaul (Freeman 2006). Hornsey also speculates on historical aspects of the development of palm wine, fermented tree sap (e.g., birch, maple), and the *kvass* (a low-alcohol-content beverage produced via yeast and lactic acid bacteria) of eastern Europe. All are ancient in origin. Very early, preagricultural familiarity with "magic mushrooms" may have been the impetus for experimentation with fermentation. Mushrooms are seasonal, whereas dried fruits, tree sap, and (eventually, with agriculture) grain could be more reliably obtained and, with some experience, fermented (Hornsey 2003). Hornsey gives a capsule history of early brewing, including pertinent basic chemistry and principles of fermentation.

Dineley and Dineley (2000) present a detailed exposition of ale and sweet malt production in Neolithic Britain. The Dineleys describe their home as "in effect, a small domestic brewery." Protocols and equipment are described suitable for both purists (those wishing to make their own quernstones) and persons willing to resort to handy plastic tubes and buckets. (Replication of ancient beers is apparently becoming popular. In addition to Ishida's and Scheer's beers above, analogous exercises in recreating ancient beverages include "Heather Ale" based on Druidic beverages, "Tutankhamen Ale," and a rice beer from Neolithic China, all referenced in Gallagher [2005]: "I proved to myself that it was drinkable and caught enough of a buzz to trip over the border in my garden. I could hear my Stone Age fraternity brothers cheering.") Dineley and Dineley note strong similarities between vessels from Britain (presumably used in production and consumption of ales) and those from specified cultural contexts on mainland Europe. Like Hornsey (2003), they refer to the question of "primacy of beer over bread"; i.e., whether techniques for beverage fermentation preceded or followed those for baking, and provide references for the debate.

In reviewing practices in Celtic Europe, Wood (2000) notes the tendency of grain placed at the edges of storage pits to sprout and suggests that preservation of this germinated grain by drying produced the first malted grain used in beverage production in Europe. Fernández-Armesto (2002) even suggests that beer production may have been the primary impetus for domestication of cereals, especially in the context of generating a surplus for "chieftainly feasts." Another radical hypothesis is that "the oldest domesticated plant may be yeast" (Braidwood et al. 1953). At any rate, brewing of beer and mead eventually became a well-established part of northern European culture, as did importation of wine.

The image of the beer- or wine-swilling barbarian has a basis in reality. According to Tacitus in *Germania* (Mattingly's [1967] translation), "Drinking bouts, lasting a day and a night, are not considered in any way disgraceful. Such quarrels as inevitably arise over the cups are seldom settled by mere hard words, but by blows and wounds. Nonetheless, they often make banquets an occasion for discussing such serious affairs as the reconciliation of enemies, the forming of marriage alliances, the adoption of new chiefs, and even the choice of peace or war." The Celts also mixed feasting, drinking, governance, and brawling, and Celtic fondness for wine was probably responsible for at least one of their defeats in battle. After successfully fighting their way through the Balkans, Celtic hordes commanded by Brennus attacked Delphi in 279 B.C., but their performance was decidedly suboptimal due to excessive consumption of Greek wine the night before the battle. Their defeat was so thorough that Brennus (after drinking still more wine) killed himself with his own dagger (Freeman 2006).

The geographic origin of wine production is uncertain, but residues of ancient wine have been identified in Neolithic pottery from Godin Tepe in the Zagros Mountains, and there is also very early evidence from the Neolithic Caucasus (McGovern 2003). Wine is unequivocally mentioned in cuneiform tablets from Ur of the twenty-third century B.C., but the ideogram for grape occurs hundreds of years before this time. The Ur tablets state that wine was brought from the mountains in great urns. Wine is specifically denoted in hieroglyphics from the first and second dynasties of the Egyptian Early Dynastic, approximately 3100–2700 B.C. The Egyptians were already employing sophisticated viticulture in the Early Dynastic period, including the use of trellises and wine presses, and were labeling wine by vintage (McGovern 2003).

Sequences of ribosomal DNA amplified from Egyptian wine jars dated ca. 3150 B.C. correspond to sequences in *S. cerevisiae* (Cavalieri et al. 2003). Kislev (1982) also cites reports of yeast cells detected in remnants of wine and beer from ancient Egypt. The extent to which *S. cerevisiae* represents domesticated rather than wild yeast, and whether it was first domesticated for wine or beer (or bread), have been considerably debated. Fay and Benavides (2005) summarize some of the literature, compare diversity within genomes of domesticated and wild *S. cerevisiae*, and conclude that strains specialized for alcohol production were less diverse and were derived from wild strains.

Although off to a slow start compared to beer, wine eventually became a major component of trade in Mesopotamia and was successfully produced in Assyria. There is no evidence that wine was used in cooking in ancient Mesopotamia, but there were multiple recognized types of wines. Egypt began importing wine from the southern Levant in the Early Dynastic. McGovern (2003) describes the extent and practice of preserving and flavoring such wine with resins. Resinated wine was the norm for much of early antiquity; resin of the terebinth tree (usually *Pistacia atlantica*, but sometimes *P. lentiscus*) was employed most frequently but was eventually replaced by pine resins. The retsina at your local Greek restaurant is a descendant of such vintages.

Summaries of techniques used in wine production in Egypt and the ancient Near East (Darby et al. 1977, Jennings et al. 2005) make it apparent that there were many distinct vintages, usually designated by place of production and varying in age, color, and sweetness. Some had special additives, and others were made from fig, palm, dates, and pomegranate. Although these wines kept better than beer, the porosity of ancient jars and the growth of undesirable microorganisms decreed that most wine be drunk young. Most wine was consumed within a year of production, although later, in Roman antiquity, "wine could be stored for many years" (Paterson 1982).

Both wine and beer are strongly reflected in ancient Near Eastern and Aegean mythology and legend. Wine was long perceived as foreign in Mesopotamia and was not accorded the high status of beer, the latter being personified as the god Ninkasi (Bottéro 2004). However, there were minor wine deities, one of which was Geštin-anna ("celestial vine"). This goddess was the sister of Damuzi (the Babylonian Tammuz), a major deity who was a shepherd but who also personified the power in grain and beer (Jacobsen 1976). Tammuz was one of several regional deities whose attributes involved fermented beverages and who suffered death and resurrection. Osiris, mentioned above, and Tammuz were analogous to the Phrygian Attis and the Syrian Adonis in that each was "a god who annually died and rose again from the dead . . . a god of many names, but essentially of one nature" (Frazer 1922; see also Frazer 1955c). Dionysius also endured death and rebirth (Otto 1965). Halperin (1983) and Saggs (2005) devote a few paragraphs to reviewing differences of opinion among scholars on the attributes of several of these deities.

In addition to personification as the Mesopotamian gods Ninkasi and Geštin-anna, fermented beverages had a major role in the Gilgamesh epic. Offerings of beer (and loaves of bread) were used by the courtesan Shamhat to civilize and humanize the wild man Enkidu (Mitchell 2004). The Egyptian goddess Hathor was associated with wine, and wine was used in revelries performed in her honor (McGovern 2003). In Greek mythology, the daughters of Aulis were gifted with the ability to turn substances into wine (or oil or grain) and helped provision the Greek fleet for its voyage to Troy. When Agamemnon became greedy and tried to make them virtual prisoners, Dionysius (see below) turned them into doves (Graves 1960). Many readers will be familiar with John 2:11 in which Jesus turns water into wine. In all, there are about 50 direct references to wine in the Bible, including those in Apocrypha, but beer drinkers will be disappointed at the lack of Biblical references to their beverage. Israel did not go without beer, but "beer did not attract the literary mind as much as wine" (Sasson 1994; see Sasson for scholarly and humorous explications of multiple Old Testament references to wine).

Special mention needs to be made of Dionysius, the most well-known deity associated with wine, who came to prominence with the rise of the Greeks. Dionysius was credited by the Greeks for bringing wine to Egypt, but given the greater antiquity of Egyptian civilization, this was a Greek conceit. Nonetheless, circumstantial evidence, especially the very early trade in drinking paraphernalia, indicates that wine was a developed commodity in the Aegean ca. 2500–2000 B.C. (Şahoğlu 2005). The mythologist Robert Graves was in rough agreement with archaeologists in placement of early grape cultivation: on the southern coast of the Black Sea by mythology, versus in the Caucasus by archaeology (Graves 1960, McGovern 2003). As noted, brewing of beer preceded wine production, and Dionysius may have evolved from a beer deity (Graves 1960, citing Harrison 1903). Greek tragedy later originated from a spring festival held in honor of Dionysius at Athens, and the rituals of this festival had a lasting impact on Greek philosophy, especially that of Plato, who had much to say about appropriate and inappropriate drunkenness (Belfiore 1986). Such spring festivals, allied with viticulture and charged with eroticism, were also part of Near Eastern culture and religion (see Gaster 1946 on girls with lips sweet as grapes, sucking nipples of the goddess, kissing and conceiving, wine for quaffing, etc.—quite a party!).

Graves (1960) recounts the travels of Dionysius and the spread of wine production as told by Greek mythology. Gimbutas (1982), like Graves, credits the deity ("Dionysus" in her spelling) with great antiquity (pre–Indo-European): "Discussions about the origin of the Greek Dionysus—whether he came to Greece from Thrace, Crete or western Asia Minor—are pointless, since all these lands originally belonged to the same Mother Culture."[2]

Bacchus was the name preferred by the Romans for their god of wine, who was identified with Dionysius. Dalby (2004) provides an entertaining if somewhat synthetic "biography" of Bacchus. The Roman god Liber Pater was also identified with Dionysius and was a god of wine, seeds, and fertility. Festivals in his honor featured a phallus paraded through fields and town and the singing of crude, rustic songs (North 2003). See also Otto (1965) for extended analysis of the Dionysian cult from a more philosophical and theological perspective. There is an extensive literature on Dionysius, Bacchus, and their roles in

---

[2] Gimbutas's stance on matriarchy was not in the mainstream of contemporary academic anthropology, but it is true that evidence for the worship of high-ranking female deities is very strong for some regions, e.g., Crete. Lapatin (2002), whose account is a well-documented and protracted tribute to skepticism in such matters, notes the importance of a female deity or deities in Minoan culture.

Greco-Roman religion and ritual. Harrison (1903, 1962) and, more concisely, Nilsson (1949) are well-regarded sources.

Frazer (1955a,b) makes interesting comments on the marriage ceremony between Dionysius and the queen at Athens, an annual ritual for fertility of vines and fruit trees, and, citing Pausanias, notes that at Potniae (near Thebes), human sacrifice had occurred in Dionysian rites prior to the substitution of goats for human victims. Graves (1960), in connection with the myth of Omphale and Heracles, notes "the vine-dresser's custom of seizing and killing a stranger at the vintage season, in honor of the Vine-spirit." Heracles was generally credited with ending such sacrifices, and even when the wars among the Greeks were at their most vicious, Greeks of the classical age took pride in the absence of human sacrifice.

More northerly peoples also had deities connected with fermented beverages, but because they came late to the use of written language, we generally know less about these deities. "The Celts had Braciaca as their god of malt and intoxication. The Teutonic people had Gambrinus, a mythical Flemish king who was said to have invented hopped beer" (Andrews 2000). Other pertinent northern gods were the Norse god Odin, who had a beer hall in Valhalla; the Norse sea god Aegir, who brewed beer in an underwater hall; and the Finnish sky god Ukko, who received libations of beer and other fermented beverages (Andrews 2000). The Celtic Goibniu hosted an underworld feast in which his beer effectively bestowed immortality on the underworld's denizens. Mead also was associated with specific myths and characters, including one myth in which it was produced from the udder of Odin's goat, Heidrun, who grazed upon Yggdrasil, the world tree of life. Drugged mead, inducing forgetfulness, was administered to the hero Sigurd in the Volsung sagas, the Norse equivalent of the Nibelungenlied. Another myth tells that mead was first brewed by dwarfs, but via an elaborate chain of contests, killing, seduction, and magical transformations became available to giants and gods (Andrews 2000). At one point, the blood of a wise (but apparently not wise enough) old man, Kvasir, is converted to one of the ingredients of the primordial mead by murderous dwarfs. Kvasir shares the same root as *kvass,* the alcoholic beverage mentioned above (Andrews 2000, Sturtevant 1952). Tales involving beer (and wine) occur sporadically throughout miscellaneous European and other folktales (usually of uncertain temporal origin), and several motifs are indexed by Uther (2004).

There are voluminous writings on all the Near Eastern and Greek deities, the circumstances of sacrifice and resurrection, and their connection to agricultural fertility and, most important from our perspective, to fermented beverages. Frazer's (1922) abridgment of his multivolume work is probably the most famous example and is still in print and widely available: "Thus the drinking of wine in the rites of a vine-god like Dionysus is not an act of revelry, it is a solemn sacrament. Yet a time comes when reasonable men find it hard to understand how any one in his senses can suppose that by eating bread or drinking wine he consumes the body of a deity. 'When we call corn Ceres and wine Bacchus,' says Cicero, 'we use a common figure of speech; but do you imagine that anybody is so insane as to believe that the thing he feeds upon is a god?'" From our point of view, however, we need not believe or disbelieve in such sacred transformations, but only acknowledge that via its metabolic capabilities, yeast (in all probability, mostly *S. cerevisiae*) has manifested itself through millennia of religion and ritual!

Wine kept far better than beer (indeed, Pliny's *Historia Naturalis* mentions wines over 200 years old!) and ultimately became a staple of trade in the ancient world. Dalby (1996) sketches the growth of the wine trade in the Mediterranean and beyond, and Rostovtzeff (1941, 1957) provides multiple examples of commerce in wine and its transport. A detailed view of wine production in Italy from the Punic Wars to the third century A.D. is given by Purcell (1985), who covers a wide spectrum of the economic aspects of ancient viticulture, including problems of seasonal labor requirements, variations in quality, overseas competition, and regulation of drinking establishments. It was essential for Germanic leaders to provide generous entertainments for large retinues, with the concomitant necessity for large amounts of wine imported from the Romans (Heather 2006). "Wine was to the Germans what beer seems to have been to the Roman boatmen on the Rhine: something they just had to have in regular supply. So important was this trade that Germanic languages borrow the Latin *caupo* for wine handler" (Burns 2003). The Celts imported Greek, Etruscan, and Roman wine and apparently invented the wooden barrel for this purpose, barrels being less susceptible to breakage than amphorae (Freeman 2006). See Toussaint-Samat (1992) for a relief carving showing such barrels during transport.

The role of amphorae in transport and storage merits specific mention, as much of what is known about wine and the wine trade in antiquity comes from the study of these jars. The archaeological literature on amphorae is vast, but some important conclusions can be noted here. Their packaging and transport, shape, construction, shipping, sealing, labeling, and shelf life have been studied. "Aged wine was valued by the Romans more than [by] Greeks; some southern Italian wines were said to require 20–25 years before they were drinkable" (Twede 2002). (One experiment in duplicating the ancient wine of Pompeii, said by Pliny to require 10 years for maturation, gave disappointing results after 2 years' maturation: "Perhaps it is still a bit young," said one drinker [Johnston 2003].) Resin from pine trees was often used to line the interior of wine amphorae with a layer 1 or 2 mm thick, which of course imparted a resinous flavor (Twede 2002). Contents of amphorae (usually wine or oil), ownership, and patterns of trade can often be deduced from the amphora label and/or the mode of construction (Callender 1965, Paterson 1982). Wine was matured in large jars (dolia), then transferred to amphorae, usually provided by the buyer for transport and shipment (Paterson 1982). Much of what is known of amphorae comes from excavation of ancient shipwrecks, a catalog of which is provided by Parker (1992).

Wine in Feasting and Daily Life in Greece and Rome

The importance of wine in Greek gastronomy and culture has been extensively documented from written sources (Homeric, classical, Hellenistic, Byzantine) and pictures on Greek ceramics (Dalby 1996). Essential points include the dilution of wine with water (and the proper ritual for mixing the two), qualities of various types of wine (characterized by locale or type of manufacture), and the evolution in Greek history of the place of wine at the Greek table. Dalby attributes to Dioscorides the earliest surviving instructions for flavoring wine with pine resin, but as we have seen, addition of resins to wine did not originate with the Greeks.

The rites of wine drinking were foundational to our notions of civilized, formal discourse. For example, "the word *symposium*, often translated as 'banquet,' really means a drinking party . . . [at which] ideas as well as toasts were exchanged. Today the word *symposium* (really a barbarism: a hybridized Latinate version) is most usually applied in English to . . . a conference, . . . collection of opinions, . . . or series of articles" (Toussaint-Samat 1992). Stages of a typical Greek banquet included conversation, foods, a ceremonial libation (the *agape*, conveying love or friendship), drinking from the *psycter* (a large drinking cup, passed person to person), and mixing of wine with water (two-fifths wine with three-fifths water) in the *crater*, a large bronze mixing bowl (Toussaint-Samat 1992).

Wine was no less important to the Romans, including the Romanized citizens of the provinces. Carcopino (1940) describes wine consumption at Roman banquets, including types of wine, timing of libations, and the role of the *commissatio* (the postdinner drinking match). An amphora with wine was provided with a label (*pittacium*) indicating the vintage, and wine was mixed with water in the *cratera*, from which the drinking bowls were filled: "We may well wonder how the . . . steadiest of heads could weather the abuses of the *commissationes*!" Many of these same aspects are covered in detail by Toussaint-Samat (1992), accompanied by miscellaneous anecdotes and aphorisms from Roman authors.

The central place of wine in Roman diet and ritual has been affirmed by a number of other authors, e.g. Burns (2003, p. 48):

> *Wine production . . . was crucial to the Roman table and cult practice and required special vessels. According to tradition, always with the Romans came vines and drinking vessels; one "could not" just sip wine from a water jug without appearing a country bumpkin. When Roman wine parapherna-*

lia turns up in a Celtic context, as happens early on in southern Gaul, this probably reveals more than just that the family enjoyed wine. The vast quantities of wine amphorae, for example, have raised the possibility that perhaps the Gauls not only imbibed but used wine in religious observance . . . perhaps inspired by Roman example.

Libations of wine were an important part of Greco-Roman religious practice, e.g., the figure of Apollo pouring a libation (Cruse 2004). Analogous libations had long been integral to ritual in the ancient Near East, e.g., Ashurbanipal pouring a libation (Saggs 2005). (The uses of wine and beer in feasting and ritual are discussed more fully in "Fermented beverages and social structure.")

A few words should be given to the importance of ritual drinking vessels. Many were costly and elaborate, others simpler, and the subject is well documented in archaeological literature and in museum collections, e.g., Eluère (1998) and Marthari (1998). McGovern (2003) includes color photographs of some of the most impressive, showing drinking scenes from a Mesopotamian cylinder seal, a spouted Phrygian jug, a spectacular steatite bull rhyton from Knossos, and a bronze, ram-headed situla from Phrygia. Freeman (2006) presents photographs of finely wrought Celtic drinking flagons, the Basse-Yutz flagons of the La Tène period, also illustrated in Herm (1975) and Laing (1979). Drinking vessels in early Anglo-Saxon culture (chalices, drinking horns, clawbeakers, and others) were elaborate and distinctive (Pollington 2003). Scythians and Celts made drinking cups from the skulls of their enemies (Freeman 2006). Even more impressive, because of its literary and potential historical implications, is "Nestor's cup," excavated by Heinrich Schliemann from a shaft grave at Mycenae. This cup is so named because of its striking resemblance in detail, motif, and construction to Nestor's cup in book XI of the *Iliad* (Wood 1985). A detailed image of Schliemann's find in Robbins (2001) clearly shows the characters (golden doves, handles, twin supports) described for Nestor's cup in the *Iliad*. Homer's *Iliad*, although composed in its present form sometime in the Iron Age, reflects much of Mycenaean society and warfare (Thomas and Conant 2005), as Schliemann's finds demonstrated. Scholars increasingly accept that excavated Hittite archives contain multiple references to Mycenaean Greeks, including references to their meddling in the affairs of the Levant coast, Asia Minor, and even Troy itself (Latacz 2004). There is corresponding acceptance that the *Iliad* contains elements of historical fact, including an attack on Troy by Mycenaeans (Latacz 2004, Thomas and Conant 2005). Perhaps Nestor's cup should be seen as an archaeological manifestation of *in vino veritas*! Drinking cups of a style originating with the Mycenaeans were traded over a wide area in Bronze Age Europe and were often used as grave goods to represent the status of the departed warrior (Eluère 1998, Jensen 1998).

Edible Fungi

Although there are some plausible references to truffles or other fungi in ancient Near Eastern sources, substantial surviving documentation on edible mushrooms essentially starts with the writings of the Greeks and Romans. The Greeks were much more circumspect about dining on fungi than were the Romans. Such was the place of fungi in the Roman diet that they were the subject not only of sauces and recipes, but also of satire and invective regarding that most Roman of all vices, gluttony. Thanks largely to the efforts of scholars in the nineteenth and early twentieth centuries, we know with some certainty several of the specific fungi esteemed by Roman gourmets, including truffles, *Boletus* species (in all probability *B. edulis*), and *Amanita caesarea* (an edible species from a very dangerous genus). The names applied to different groups of fungi by the Greeks and Romans were used by the great classifier Linnaeus without respect to their original meanings, so modern readers must beware when a Roman author writes of *boletus* or *pezieae*, as these words had quite different implications for ancient readers than they do for modern ones.

Europeans north of the Mediterranean seem to have been even more divided than Greeks and Romans on the merits of mycophagy. Unfortunately, those Europeans came late to literacy, so details of their mushroom-eating habits, or lack thereof, must usually be inferred on the basis of folklore or later habits. However, there is much evidence from both folklore and modern habits to indicate that eastern European (especially Slavic) peoples have long entertained a taste for mushrooms, whereas most western Europeans, especially the British and the Dutch, have had less enthusiasm for fungi at the dinner table. Some persons with a decidedly mycocentric view of culture have divided entire societies into mycophilic versus mycophobic. See also Fungi in Ancient European Folklore.

Documentation: Edible Fungi

Mycophagy in Greece and Rome

Explicit information on mycophagy in early Greco-Roman antiquity is difficult to locate. Theophrastus (ca. 300 B.C.) was the first Greek writer to mention fungi as a sort of organism. Other Greeks who wrote about fungi include Nicander (ca. 185 B.C.) and, much later, Dioscorides (second century AD). Unlike the Romans, the Greeks were not so fond of fungi: "The Greeks have always been mycophobic" (Wasson and Wasson 1957). In contrast, Ovid, Horace, Celsus, and other Roman writers made frequent references to fungi, including many references to fungi as food. Plutarch, a Greek, also wrote about fungi in the diet.

The most extensive scholarly commentary on Greco-Roman mycology and mycophagy is probably Buller's (1915) presidential address to the British Mycological Society, and before that, Houghton (1885). Buller listed additional references from nineteenth- and early twentieth-century scholarship. The most striking superficial difference between ancient and modern mycology is nomenclatural and derives from the misuse of classical names by Linnaeus. Readers who have even a superficial acquaintance with modern names for common fungi will note with dismay that the Roman *boletus* corresponds to *A. caesarea*, the Roman *agaricum* to *Polyporus officinalis* (current name *Laricifomes officinalis*), and the Roman *pezieae* to puffballs in *Bovista* and *Lycoperdon*. Thus, Linnaeus used the Roman name for a gilled mushroom (*Amanita*) for a genus accommodating mushrooms with pores instead of gills. He used the name for a polypore to establish a genus for mushrooms with gills, and the Roman name for puffballs was applied to cup fungi. Linnaeus also used the Greek ὕδνον (for what we now call truffle) as the basis for *Hydnum*, fungi with a toothed hymenium. Houghton (1885) and Buller (1915) provided further examples of how the otherwise astute Linnaeus ignored the use and context of the ancient names when erecting his fungal genera. Ainsworth (1976) renders a concise list of literature on the subject but concludes, "No serious attempt has been made to check or extend the references to fungi in the Greek and Roman classics" compiled by Buller and miscellaneous nineteenth-century authors. Grivetti (2001) still cited Houghton (1885) as the principal source of such references. However, there have been concerted etymological studies of words used to name a few of the pertinent genera, e.g. *Boletus, Helvella,* and *Suillus,* with miscellaneous points of contention remaining (Imholtz 1977).

Once the nomenclatural disjunction is assimilated, one can readily apprehend the impact of fungi on Greek, and more especially Roman, cuisine. The ancients dined on truffles, edible species of *Amanita,* other mushrooms including *Boletus edulis* and allied species, and puffballs (Buller 1915, Houghton 1885). Identification of other edible fungi mentioned by classical writers is more speculative. Strong possibilities include the meadow mushroom *Agaricus campestris*, the fairy ring mushroom *Marasmius oreades*, *Russula alutacea*, the shaggy mane *Coprinus comatus* (Houghton 1885), species of *Pleurotus*, and very probably *Lactarius deliciosus* (from a mural at Pompeii) (Buller 1915). Buller also mentioned as possibilities true or false morels, chanterelles, *Hydnum repandum* (the hedgehog fungus), and *Fistulina hepatica* (the beefsteak fungus). It is possible that Greeks and Romans knew how to cultivate the poplar fungus, *Pholiota aegerita* (*Agrocybe aegerita*), by mixing poplar bark with dung or by treating poplar stumps with proper moisture and nutrients (Buller 1915, Houghton 1885). *Terfezia* spp. and *Tirmania* spp., trufflelike ascomycetes also known as desert truffles or poor man's truffles, were imported in quantity from northern Africa to Rome (Pegler 2002). And "fungus stones" (*Polyporus tuberaster*) were "well known and much prized in Ancient Rome, where they were sold under the name 'Lapideus,' and continued to be traded throughout . . . the Italian Renaissance" (Pegler 2000a).

Several Roman writers (Pliny, Juvenal, Martial, Cicero) and at least one Greek (Plutarch) occasionally made acerbic comments on the habits of voluptuaries of their time by mocking the content or preparation of dishes containing fungi. Buller (1915) translated some of the more humorous epigrams, including one from Martial, who has a cooking pot lament that it was formerly used for mushrooms, but "I am now used, I am ashamed to say, for Brussels sprouts." Buller also recounted Martial's opinion that messengers who could be trusted with gold and silver might succumb to temptation if allowed to transport mushrooms. Carcopino (1940), in his popular work on daily life in Rome, repeats comments by Juvenal and others on truffles and gluttony. St. Augustine's reference (in *De moribus Manichaeorum*) to a belching Manichaean, eating mushrooms until his stomach is distended, can be taken to be in this same tradition of equating gluttony and mycophagy (but for a more risqué interpretation, see footnote 122 in Ruck et al. 2001). Toussaint-Samat (1992) also notes St. Augustine's equation of truffles, mushrooms, and gluttony when the noted philosopher took aim at the Manichaeans. Toussaint-Samat (1992) relates more anecdotes about truffles from the writings of the Greeks and Romans as a prelude to describing their roles in medieval and Renaissance cuisine. One singularly interesting account is that of a metic (resident alien) of Athens, who was awarded citizenship in return for a dish of truffles.

Recipes for preparing fungi are rarer than documentation of fungi as food, but we have some references for mushroom preparation from the Roman tradition. The Roman gourmet Apicius listed some ingredients used with truffles, *boleti,* and (probably) morels. Prominent in one or more recipes are wine or wine vinegar, salt, pepper, oil, coriander, mint, cumin, parsley, and possibly leeks (Houghton 1885). Solomon's (1995) review of ancient Roman cooking, "The Apician Sauce," pays scant attention to fungi and says, in seeming contradiction to Houghton (1885), that recipes for mushrooms were simpler than for other food items. Nonetheless, fungi are cited in the review in the context of an extended description of the flavoring of Roman dishes. Solomon describes in detail the uses of wine and wine vinegar, products of fungi to be sure, in a variety of sauces. Soyer's (1853) *Pantropheon* included mushrooms and truffles in the chapter on seasonings (and had sections on wine and drinking cups as well). Occasionally one encounters a recipe that seems to step over the line. Even the most dedicated of modern mycophagists might balk at the practice of garnishing their mushroom dishes with pearls (Faas 2003), a form of mycodecadence with no present-day equivalent.

Dalby's (1996) compendium of gastronomy in ancient Greece merely mentions mushrooms, but again wine and wine vinegar are prominently featured. Perhaps this is because "Greeks were not very enthusiastic about mushrooms" (Brothwell and Brothwell 1969). Braun (1995) cites two Greek classical sources, Poliochos and Antiphanes, to the effect that destitute folk would eat wild plants such as sow thistle and mushrooms. That Roman tastes were decidedly more mycophilic is documented by the several sources discussed in detail by Buller (1915) and Houghton (1885) above.

Mycophagy in Other Cultures

Use of mushrooms in cuisine is more difficult to trace in cultures of antiquity other than Greco-Roman, probably because macro-fungi are rare in the drier climates of Egypt and west Asia, and because other cultures with more suitable climate (western and northern Europe) lacked systems of writing appropriate for recording such information. Bottéro's (2004) review of Mesopotamian cuisine does not mention mushrooms, although letters excavated from Mari mention truffles (Brothwell and Brothwell 1969), and Sumerian writings include observations on the Amorites, nomads in the mountains, "who dig for truffles" (Saggs 2005). The two-volume work of Darby et al. (1977) on food in ancient Egypt does not address mushrooms in the Egyptian diet, but apparently the ancient Egyptians at one time forbade the consumption of mushrooms by commoners (Andrews 2000, Kavaler 1965).

Mycophagy in ancient Europe north of the Mediterranean is also difficult to infer. However, Salomonsson (1990) draws attention to the general prejudice against eating fungi in northern and western Europe, in contrast to its general acceptance in southern and eastern Europe, and further notes that "taboos and prejudices against certain kinds of foods can be of extremely long duration." Much has been made of the supposedly mycophobic versus mycophilic attitudes of various cultures. Evidence from herbals, poetry, fiction, and culinary writings supports the notion that Anglo-Saxons in general are averse to eating fungi, in contrast to the mycophilic tendencies of continental Europeans (except in Holland) and especially Russians (Benjamin 1995). "Wild mushroom consumption is popular throughout most of Europe, except Britain and the Low Countries," according to Morgan (1995). Wasson and Wasson (1957) strongly support the idea that cultures may be classified as mycophobic or mycophilic. They regard Russians (and Slavic peoples in general) as mycophiles: "Muscovy and Catalonia may be taken as the citadels of mycophagy." In contrast, "so far as to mycophobia, the foci of infection are found on the one hand among the Celts and Frisians along the shores of the Atlantic and the North Sea, and on the other hand in Greece." These conclusions are not universally accepted, at least not in their radical form (Vogt 1958). Nonetheless, these general patterns show some conformity to those deduced from Fungi in Ancient European Folklore (Chapter 13), and they may reflect ancient culinary habits.

Fungi as "Entheogens"

Many fungi produce psychoactive metabolites (other than ethanol), and much has been written on the evidence for the use of such fungi in the context of history or myth. Several modern analyses originate from Europeans or Americans whose cultural outlook was acquired during, or at least influenced by, the experimental atmosphere of the 1960s and later, and some care must be taken to disentangle plausible events in historical antiquity from the enthusiasms and experiences of the modern authors. Nonetheless, several such authors have established reputations as highly competent scholars in linguistics, mycology, mythology, and/or ancient history. Their arguments must be taken seriously, even when their conclusions strongly contradict current conventional religious beliefs or orthodox historical perspectives. They have proposed several interconnected hypotheses, which may be roughly stated as follows: (1) that the mushroom *Amanita muscaria* was the "soma" of the Vedic scriptures; (2) that it and/or fungal plant pathogens or endophytes growing in association with grass family plants were integral to the Eleusinian mysteries in Greece; and (3) that modern religions, most specifically Christianity, have their roots in experiences with the ingestion of psychoactive fungi. These hypotheses have acquired a life of their own in post–1960s culture, including numerous Web pages, blogs, books, etc., and can be said to some degree to form a part of modern, "invented" paganism. However, there is nothing inherently implausible about the original hypotheses, and readers can decide for themselves the degree to which they agree with the foregoing concepts and conclusions.

Documentation: Fungi as "Entheogens"

Presented here are instances of the ancient use of fungi that produce psychoactive metabolites (other than ethanol) that have been documented or at least shown plausible in the literature. Some authors have referred to these metabolites as psychoactive, others as hallucinogenic, and some have termed them "entheogens," a term adopted here because it connotes the use of psychoactive substances in the context of religion or ritual. Introduced in the late 1970s, *entheogen* affected the lexicon of ethnobotany somewhat later (Wasson 1980, Wasson et al. 1986). Although the use of fungi as inebriants in the Old World is documented in various sources, care is essential in disentangling solid evidence of historical use from modern enthusiasms and embellishments. Much substantial scholarship pertains to mycophagy and mycolatry in the Americas (e.g., Schultes 1998; Wasson 1980) and is beyond the geographic scope of this review.

Fungi, Philology, and Mythology

The most widely read and spectacular accounts are probably Wasson (1968) and Allegro (1970). Both works argue from largely philological grounds, although Wasson's account is buttressed by anthropological data on ingestion and excretion of *A. muscaria* intoxicants by shamans in Siberia. Wasson (1968) reviews the features of the divine substance soma, as it was described in the Vedic scriptures of ancient India, and descriptions of its use. He concludes that the morphology and development of *A. muscaria* strongly conform to the Vedic descriptions of soma. The long-puzzling Vedic description of soma being passed as urine was placed in the context of analogous Siberian practices, in which the intoxicant may be passed along to other users in urine of the primary consumer. Wasson also describes some evidence for much later use of *A. muscaria* in northern Europe. Wasson (1972) reiterates the soma hypothesis in abbreviated form in a volume on the worldwide historical use of hallucinogens in religion and ritual.

Wasson's theory gained wide acceptance (e.g., Lévi-Strauss 1970, Toporov 1985) and is given credence by some modern mycologists (Michelot and Melendez-Howell 2003). Consensus is far from universal, however, and other scholars have offered alternative theories regarding soma (reviewed by Merlin 2003 and Riedlinger 1993). Smith (1972) summarizes the early confirming, noncommittal, and rejecting reviews and briefly discusses disputed points.

Ruck and Staples (1994) expand the preceding hypotheses by explaining that the Mycenaean Greeks were Indo-European invaders who brought their use of *A. muscaria* with them into archaic Greece and combined Indo-European practices with the aboriginal inhabitants' pre-existing use of intoxicants. Ruck (1983) argues that the secret "offerings of the Hyperboreans" that were presented, wrapped in straw, at Delos may have been dried *A. muscaria*. He argues at length on the basis of philology and analysis of mythology that these offerings were integral to an understanding of the transformation of ancient Greek religion from primitive to classical forms. Littleton (1986) considers the possibility that *A. muscaria* or other psychoactive fungi were responsible for the intoxicating vapors (*pneuma enthusiastikon*) at the oracle of Delphi but opts for *Cannabis* as a more plausible cause.

Allegro's (1970) account has garnered less acceptance, no doubt in part because his conclusions obtrude on religious sensibilities. He relies on intricate philological analysis of Sumerian, Akkadian, Arabic, Aramaic, Greek, Hebrew, Latin, Egyptian, Persian, Sanskrit, Ugaritic, and other languages. His status as a translator and analyst of the Dead Sea Scrolls lends his arguments the weight of authority. Persons not highly knowledgeable in half a dozen ancient Near Eastern languages will find it difficult to fault specifics of Allegro's philological reasoning, although they may not accept his final conclusions. Essentially, Allegro argues that ritual use of psychoactive fungi, specifically *A. muscaria*, was the underlying basis for much religious experience in ancient Iran, Mesopotamia, and adjacent areas, including Palestine, and that Christianity was one of the products of this process. His inclusion of such topics as sacred prostitution and his insistence that sexual allusions were integral to ancient religion and primitive Christianity have not endeared his theories to traditionalists. King (1970) gives a detailed Christian critique of "the mushroom myth." The thesis was rejected on other grounds also; at least one review of fungi in folklore discredits Allegro's conclusions on the basis of purported geographic distribution of the mushroom (Findlay 1982). Although *A. muscaria* is a worldwide species widely distributed in Asia and the Mediterranean region, conclusive documentation of its occurrence in the Levant was lacking in a world survey (Guzmán et al. 1998).

The poet, novelist, and scholar Robert Graves was a convert to the idea that *A. muscaria* had a huge impact on ancient religiosity. Graves (1960) introduces his two-volume work on Greek mythology with reference to the ritual use of *A. muscaria* and *Panaeolus papilionaceus* in remote antiquity. The cover to the 1979 Penguin reprint of *The Greek Myths* (vol. 1) shows a fifth-century B.C. Greek relief carving of two individuals holding aloft two mushrooms, both of which have stipes (stems) with swollen volvas (bases) and one of which seems to have a faint annulus (ring) on the stipe, structures typical of *Amanita* species. However, Samorini and Camilla (1994), when illustrating and analyzing the same carving, mention *Psilocybe* or *Panaeolus* as candidates for the mushrooms. The two individuals were said to be Demeter and Persephone. Ruck et al. (2001) present an interesting and detailed analysis of the carving, but the reproduction itself in their volume is inferior in quality to that on the Penguin reprint of Graves (1960). (See Figure 10 in Ancient Images of Fungi.)

Graves (1958), exploring the topic in greater detail in his chapter "What Food the Centaurs Ate," uses extended philological arguments to conclude that the ambrosia consumed by the Greek deities contained fungi, as did the potions consumed in the Eleusinian mysteries. Perhaps the fly agaric (*A. muscaria*) was consumed by the Maenads. Perhaps the feasts of Dionysius (termed "the Ambrosia") were originally mushroom orgies. Graves entertains these and other speculations. Among the most imaginative is the idea that Alexander's conquest of Asia was in emulation of the mythical Dionysius, who in deepest antiquity had invaded Asia Minor and destroyed armies of Persians and who had gone as far as Bactria. According to Graves, the myth of the Dionysian invasion of Asia (a well-documented myth mentioned in Euripides) was probably founded on a raid into Thrace by *Amanita*-intoxicated warriors from Macedonia.[3] Because Alexander wanted to imitate Dionysius, he led his armies east, not west. Hence, "the fly-amanite may have changed the course of European history, and even been the reason why the common language of Britain and the States is not a barbarous dialect of Greek." In addition to such poetic inspiration and speculation, "What Food the Centaurs Ate" also has useful information on images of fungi, such as the Etruscan portrayal of a mushroom with Ixion (see Fungi Used for Medicinal Purposes and Other Technologies) and mushrooms on Greek vase paintings (see Ancient Images of Fungi).

In a work devoted largely to western European myth, Graves (1966) reiterates his ideas on *A. muscaria* as a means to Dionysian ecstasy and also points to its relevance in the connection between toads and toadstools (though not with the same diligence as Morgan [1995] or Wasson and Wasson [1957]). Graves (1966) also refers to *P. papilionaceus* as consumed in the mysteries at Eleusis and analogous mysteries in Crete and Samothrace but notes that ultimately "wine displaced toadstools at the Maenad revels." Baal Zabul (a Canaanite deity) "was an autumnal Dionysus, whose devotees intoxicated themselves on *amanita muscaria*." Graves gives the Biblical names for these toadstools as "ermrods" or "little foxes"; he does not specify chapter and verse, but if this is an allusion to Song of Solomon 2:15, the context makes "foxes" more plausible than mushrooms! At any rate, Baal Zabul became Baal-Zebub (Beelzebub), "Lord of Flies" (Graves 1966), a fitting appellation for a deity allied to the fly agaric.

In seeming imitation of Wasson and Graves, Wilson (1999) attempts to discern evidence for soma in Celtic (and most specifically, Irish) legends, but the evidence produced is highly indirect. Morgan (1995) summarizes much of the literature on soma and the fly agaric and states that "it can be assumed that the fly agaric was associated with shamanism and intoxication in the region of the Black Sea, from at least as far back as five to six thousand years ago, when the Samoyed first migrated north." This assumption is questionable, however, because the ancient movements of peoples throughout the steppe regions of Eurasia are the subject of

---

[3] Eliade (1972) reviews the history and culture of *berserkir*-like warriors in the Balkans, including the possible roles of intoxication and soma-like substances.

intense controversy. (See Lamberg-Karlovsky [2002] on this subject, as well as for references to some nonfungal candidates for the soma of Indo-Iranian peoples.)

Regardless of whether *A. muscaria* was the soma of the Vedas, it is possible that it was among the archaic hallucinogens used in Europe, including Bronze Age Scandinavia. Rudgley (1995) summarizes evidence and literature for use of hallucinogens in medieval and ancient Europe but provides little evidence for use of *A. muscaria* beyond citing Kaplan (1975). Morgan (1995) covers some of the same territory but is skeptical about the purported use of *A. muscaria* by the Viking berserkers (warriors who fought ferociously with seeming disregard for injury or pain). Fabing (1956) favors the hypothesis of *A. muscaria* use by certain Vikings as an explanation for berserker behavior and presents the literature on the subject in detail. The successful outlawing of berserkers by legislation in 1123 (the punishment was banishment) is taken by Fabing as proof that the behavior was drug-induced and preventable. If true, this would constitute one of the few instances in which drug use (or at least its associated behavior) has been successfully eliminated by legislation and threats of punishment. Neither mushrooms nor any other specific psychoactive substances are mentioned in the review of "Berserks" by Lindow (2000), who concentrates on literary sources (various Icelandic-Norse sagas) and etymologies.

Other fungi said to be significant in ancient rituals and in promotion of religious fervor were species of *Claviceps,* agents of the ergot disease of cereal grains and grasses (Figure 2) (Wasson et al. 1978). Alkaloids produced in some ergot sclerotia are the immediate precursors of LSD. Wasson et al. (1978) assert that the psychoactive properties of ergot were recognized in antiquity and that ergot of barley was used in a potion consumed in the Eleusinian mysteries. Since the ceremonies at Eleusis were secret (hence "Eleusinian mysteries"), much of the authors' argument is inherently circumstantial and philological, with occasional support from paintings on vases and fragments of commentary from ancient literature. Verster's (1976) abstract reaches similar conclusions. Ruck and Staples (1994) cover much of the same ground and include a review of the role of *Lolium temulentum* ("drunken Lolium" or darnel), a host of ergot (and sometimes of a related fungal endophyte with toxic alkaloids). Based on the Biblical descriptions, archaeological evidence, and contemporary field reports, the tares of the Bible (Matthew 13:24–30) were indeed probably *L. temulentum* (Moldenke and Moldenke 1952, Musselman 2000).

The species of *Claviceps* most closely linked with production of psychoactive compounds in modern times is *C. paspali,* which grows in the host grass *Paspalum digitatum.* Hulme (2004) states that *P. digitatum* is common in the Mediterranean and a host of *C. paspali,* but the weedy grass is native to the Americas and was presumably not present in the Mediterranean during antiquity. Ruck (1983) states that *P. distichum* was the pertinent host species and that this plant "contains pure, uncontaminated entheogenic alkaloids." Ruck gives an extremely detailed presentation on the use and significance of consciousness-altering fungi (presumably including endophytes of *P. distichum*) in ancient Greek religious practices. It is a matter of continuing debate which, if any, species of *Claviceps* were involved, and what was the most probable host (Webster et al. 2000). The views of Wasson and his colleagues regarding identification of *A. muscaria* as the soma of the Vedic scriptures have met with considerable (but not universal) acclaim, as documented above, but reception of analogous hypotheses for the importance of fungi as entheogens in ancient Greece has been more mixed (e.g., Kearns 1989; Luck 2001).

C. A. P. Ruck was one of the coauthors of Wasson et al. (1978). Ruck and Staples (1994) present an interesting ancillary hypothesis: When the invading Indo-Europeans established new social structures in Greece, they introduced a false etymology, deriving the name for Mycenae from *mykes*, which is Greek for mushroom. Ruck (1983) presents the same

FIGURE 2

Ergot *(Claviceps purpurea)* sclerotia. *C. purpurea* is one of several species of *Claviceps* infecting grass family plants and the species primarily responsible for production of highly toxic alkaloids in cereal grains. (From Engler 1897)

idea: "The ancient foundations of that city were supposedly built by the Cyclopes, and it was named, as it was claimed, after the 'mushroom' or *mykes* that Perseus picked at the site (Pausanias 2. 16. 3)." He presents philological evidence that names for Mycenae, Athens, and Thebes are pre-Hellenic (i.e., in this case, pre–Indo-European) and that the invading Indo-Europeans reinterpreted the place name of Mycenae in the context of their own sacred plant lore. Graves (1960) relates the original myth in a footnote: "Perseus . . . founded Mycenae, so called because, when he was thirsty, a mushroom [*mycos*] sprang up, and provided him with a stream of water." (Morgan [1995] and Ruck et al. [2001] reproduce an image from a Greek vase showing Perseus and mushrooms; see Ancient Images of Fungi.) Ruck et al. (2001) claim great significance for a connection between Perseus and the fly agaric (*A. muscaria*) and also claim that the golden fleece sought by the Argonaut Jason was "ultimately *Amanita muscaria.*" It is safe to say that the conclusions of Ruck et al. (2001) on the use of plant-derived drugs ("entheogens") by pagans, as well as by Jesus ("the Drug Man") and Moses, represent a minority view, albeit an entertaining one. However, they reproduce several illustrations of interest, e.g., Ixion bound to the wheel with mushroom tinder at his feet (Ruck et al. think the "tinder" was actually *Datura stramonium*), the Greek vase with Perseus and mushrooms, Persephone and Demeter with mushrooms, and several other interesting figures, including one with a mushroom and reindeer. Ruck (2006) is a very readable and well-illustrated synthesis and extension of these ideas. However, the only specific documentation is an extended series of quotations, largely from classical sources, supplemented with a list of sources (all works by Ruck and his colleagues).

The myths regarding Perseus have not gone unnoticed by professional mycologists. From the introduction in Alexopoulos (1962, p. 3):

> *Three and one-half millennia ago, so the legend goes, the Greek hero Perseus, in fulfillment of an oracle, accidentally killed his grandfather, Acrisius, whom he was to succeed on the throne of Argos. Then, according to Pausanias, "When Perseus returned to Argos, ashamed of the notoriety of the homicide, he persuaded Megapenthes, son of Proteus, to change kingdoms with him. So, when he had received the kingdom of Proteus he founded Mycenae, because there the cap (mykes) of his scabbard had fallen off, and he regarded this as a sign to found a city. I have also heard that being thirsty he chanced to take up a mushroom (mykes) and that water flowing from it he drank, and being pleased gave the place the name of Mycenae."*

Among his sources, Alexopoulos cites Frazer (1898) and Ramsbottom (1953). The latter, a mycologist, also provides a short summary of references to fungi by Greek and Roman authors, as well as in some medieval herbals and other texts. Interestingly, fungi may be involved in another classical foundation myth: The mother of Romulus and Remus, founders of Rome, was said to be impregnated by a phallic entity arising from flames and ashes of a hearth. Wasson and Wasson (1957) recount opinions regarding candidate fungi (a stinkhorn versus a morel) that might account for the myth.

Matriarchies, Tanists, and Ritual Sacrifice

One difficulty with assessing the literature on psychoactive fungi in antiquity is that authors who were foremost in documenting the evidence (e.g., Graves, Ruck, Staples) were also proponents of the concept of a widespread ancient matriarchal civilization that was supplanted (indeed, overthrown) by patriarchy. This perspective, more accepted by scholars in the early and mid-twentieth century than at present, represents a diverse tradition with numerous adherents and an extensive literature. Briffault (1931) is arguably the best example. The idea of widespread ancestral matriarchy is no longer given much credence by mainstream academic anthropologists. "There were no benign matriarchs, nor male rebellions" (Marangudakis 2004). Nonetheless, a substantial body of literature still reflects the idea of ancient matriarchy (e.g., Russell 1998), and lively academic debate persists (e.g., Berggren and Harrod 1996; Christ 2000).

Embedded in the idea of patriarchal suppression of the matriarchy are the allied concepts of patriarchal suppression of shamanism (and/or witchcraft and other spiritual and religious practices), restrictions on women's sexuality, and, eventually, prohibition of psychoactive substances (other than ethanol, nicotine, and caffeine). From a scientific perspective, there is no a priori reason that use of psychoactive fungi must have occurred in any particular social context, matriarchal, patriarchal, or otherwise; the evidence for use of psychoactive fungi has to be assessed on its own merits. Concepts of ancient matriarchy were often bound up with notions of primitive sacrificial rites (including even the sacrifice of kings or their ritual "stand-ins" known as tanists) and the fertility of vineyards or other crops. Readers need to be aware of the social and cultural contexts in which various ideas on psychoactive fungi have been presented.[4]

---

[4] Many of the views of Graves (and, indeed, of James Frazer) were bound up with notions of ritual sacrifice of kings or their tanists and the connection of these sacrifices and other rituals to maintaining fertility of fields or vineyards. Pharand (2005) gives a sympathetic but candid review of Graves's views on mythology and history. Frazer's theories on the evolution and significance of magic, folklore, and religion are now regarded as antiquated and inconsistent, but *The Golden Bough* still appeals to a wide audience and sells in large numbers (Ackerman 1975; Beard 1992). Beard hypothesizes that "the success of the *Golden Bough* rests on the undeniable fact that it is so rarely read" (a counterintuitive and paradoxical hypothesis that "does not derive from any statistically accurate survey" but is based on the assumption that persons making a pretense of literary culture want it on their bookshelves). However, Frazer's work presents "a wider range of information about religious and magical practices than has been achieved subsequently by any other single anthropologist" (Encyclopedia Britannica Online 2006). The continued relevance (and commercial success) of *The Golden Bough* lies not in the utility of Frazer's theories, but in the fact that it represents, even in the abridged version, a huge catalog of world folklore, myth, and ritual (including examples pertinent to our subject). Battey (2003a,b), for example, writing from the perspective of contemporary plant science, mines the works of Frazer and Graves for examples of the impact of vegetation on human spirituality, emotions, and aesthetics.

Poisonous Fungi and Mycotoxins

Poisonous mushrooms gained notoriety even in antiquity for their part in destroying innocent or not so innocent lives. Especially famous was the reputed poisoning of a Roman emperor. The fungal toxins, and the dishes into which the toxins may have been introduced, have been the subjects of considerable speculation. Some cultures, especially the Greeks and Anglo-Saxons, long retained a suspicion of fungi as foodstuffs. Antidotes against poisoning (mostly emetics composed of pungent substances) were probably effective primarily against early-onset gastrointestinal difficulties. They would have been of little use against the toxic *Amanita* species, whose symptoms may not appear for several hours after ingestion.

Mycotoxins were a threat in both food and forage. Much has been written in modern times regarding probable connections between mycotoxins and events in medieval, Renaissance, or colonial times, including instances of mass hallucinations, changes in birthrates, charges of witchcraft, etc. The impact of mycotoxins in Europe undoubtedly rose in the Middle Ages, with the spread of cultivation of rye, an excellent host for ergot (*Claviceps purpurea*). But ancient peoples probably also suffered from ergot as well as from the effects of *Fusarium, Penicillium,* or *Aspergillus* mycotoxins and from allergens of common fungi. Dramatic claims have also been advanced to explain several events in antiquity. Some, such as the notion that fungi turned the Etruscans into homosexuals, are best relegated to the realm of unintentional humor, but other claims bear close examination.

The ancients were well aware that foodstuffs, including grains, could become moldy, but they had no notion of microorganisms per se. To appreciate the probable impact of toxigenic microorganisms in the ancient world, it is essential to understand in some detail the norms and limits of ancient storage and transport of grains. What we know of these practices comes partly from historical and archaeological data and partly from modern experimentation that seeks to duplicate ancient storage conditions.

Ancient peoples took deliberate and elaborate measures to exclude moisture and provide adequate ventilation in their granaries because they knew that failure to do so would result in spoilage. Small-scale storage in pits may have sometimes purposely excluded air and succeeded in creating anaerobic conditions for long-term storage, but there are indications that success was mixed. Storage pits in moist climates or seasons were probably common sources of contaminated grain. Modern experiments have shown that toxigenic fungi can be recovered from such pits, and colonization of the grain by such fungi could be extensive. The British in particular have shown an enthusiasm for duplicating the storage methods of their ancestors, sometimes finding so little deterioration that some skepticism seems warranted. Recent findings may dispel modern myths about old crops; for example, the notion that covered wheat (such as spelt, whose hull is retained on the kernel at harvest) is more protected from pathogens than "naked" wheat (such as modern bread wheat, which lacks such a hull on harvested kernels) has been strongly called into question by some modern research.

Documentation: Poisonous Fungi and Mycotoxins

Poisonous Fungi

The Greeks and Romans were highly aware that some mushrooms were poisonous (Ainsworth 1986, Buller 1915, Houghton 1885). The Greeks early associated fungi with toxins (e.g., the "evil ferment" postulated by Nicander) and attempted to devise remedies for those who had ingested poisonous mushrooms. Buller relates incidents described in classical Greek literature (Eparchides, Hippocrates) concerning fatal and nonfatal poisonings by fungi. Ainsworth (1993b) credits Euripides with a reference to mycetism (eating a poisonous fungus mistaken for an edible one) killing a woman and her two children. The civilizations of preclassical and classical antiquity were not sufficiently developed scientifically to arrive at consistently useful criteria for distinguishing toxic and edible species.

Remedies were varied, including pungent mustard, bird dung, and vinegar mixed into various concoctions, several of which were effective emetics (Houghton 1885). Hippocrates and Nicander both provided an assortment of remedies against poisonous fungi, and Celsus, a contemporary of Augustus, recommended boiling fungi with the young twig of a pear tree to free them of toxic properties (Phillips 1982). Dioscorides (first century AD) thought that certain substrata that fungi grew on or near (serpents' dens, rotten rags, rusty nails, etc.) would make them poisonous and recommended as a cure the usual dose of vinegar and bird dung. Houghton (1885) traced the genesis and transmittal of such ideas throughout ancient commentaries.

Dioscorides and Galen (b. A.D. 130) had great influence on medical thought up through the Renaissance, including the writings of the herbalists of the sixteenth and seventeenth centuries (Buller 1915). On the whole, Galen took a negative view of fungi in the diet, with the possible exceptions of the meadow mushroom (modern name *Agaricus campestris*) and Caesar's mushroom (*Amanita caesarea*, which is edible but which, because it belongs to a genus with so many toxic species, is best avoided by all but the most expert mycologists). Further excerpts from or summaries of Dioscorides, Celsus, Pliny, and Galen are provided by Phillips (1982), and Buller (1915) briefly references Roman writers of the second and third centuries A.D. who documented cases of poisoning by fungi.

Let us turn now to the deliberate use of poison in dishes containing edible fungi (*boleti*), allegedly the means by which Agrippina rid herself of her troublesome husband, the emperor Tiberius Claudius. Because Roman emperors of that era were deified when deceased, Nero, the son of Agrippina and Claudius, later quipped that mushrooms were the food of the gods (Grant 1970). A similar incident at a banquet at which mushrooms (edible *suilli*) were served ended the life of Annaeus Serenus, the prefect of Nero's guard. For the death of Claudius, Graves (1958) blamed "juice of the lethal *amanita phalloides*, added to the *amanita caesarea*, an edible mushroom of which Claudius was extravagantly fond." However, some writers believe that the feast that killed Claudius was based on poisonous *Amanita* species. Molitoris (2002) seems to echo this view, and Grimm-Samuel (1991) vigorously defends it. Grant (1975) also seems to subscribe to this notion, adding that "accidental loss of life frequently occurs in Italy due to confusions between the harmless mushroom boletus edulis and the fatal amanita phalloides." But this degree of confusion seems unlikely, given that *B. edulis* has a hymenium of pores, whereas *A. phalloides* is a gilled mushroom. It is much more likely that young *A. phalloides* might be mistaken for immature edible *A. caesarea*. (The account given by Toussaint-Samat [1992] should be disregarded, as it confuses *A. phalloides* with *A. muscaria* and moreover misapplies the common name "Caesar's mushroom" to *A. muscaria*.)

Poisonous fungi were actually only a small part of the arsenal of drugs and poisons employed in ancient Rome (Cilliers and Retief 2000). And despite multiple commentaries from ancient and modern sources implicating poison, usually from mushrooms, Marmion and Wiedemann (2002) argue that Claudius was not poisoned at all, but simply suffered "sudden death from cerebrovascular disease."

Mycotoxins

Most approaches to assessing the potential impact of mycotoxins in antiquity are indirect. They involve the appraisal of ancient storage practices, experiments conducted under conditions replicating these practices, experiments with plant taxa known as ancient crops, certain findings pertinent to other microorganisms or ancient insects, and occasional references to possible symptoms of mycotoxicosis in the writings of ancient peoples. Direct mentions of mycotoxins in the writings of ancient peoples are, not surprisingly, very rare, but Galen stated that black wheat (probably smutted or rusted grain) is less harmful than ingestion of darnel (tares, a *Lolium* species infected with toxin-producing fungus (Aaronson 1989). (Fungi that produce mycotoxins when growing on building materials and fabrics are discussed in the section on rot below.)

The fungi that produce mycotoxins did not originate with storage of grain by humans and probably have an extremely ancient history in seeds stored by rodents or other animals (Hawkins 1999, Smith and Reich man 1984). However, it is likely that conditions of food storage by humans from the Neolithic onward were sometimes conducive to such fungi. This is the stated assumption of Wijbenga and Hutzinger (1984), who note the potential for chronic effects from mycotoxins in early agricultural societies but provide concrete examples of such effects only from medieval or modern periods. (Other writers have speculated freely about such effects in classical and Biblical antiquity, as documented below.) Sprouted grains from ancient stores excavated in Armenia and dated to the third millennium B.C. showed that wheat was sometimes wet going into storage or that storage conditions became moist (Gandilian 1998).

Indirect evidence that storage fungi might have invaded ancient grain stores comes from detection of tetracycline (a product of *Streptomyces* and other actinomycetes) in bone tissues recovered from ancient burials. Mills (1992) discusses the plausibility that ingestion of *Streptomyces*-contaminated grain was responsible for the tetracycline and draws inferences about grain storage conditions in ancient Egypt and Nubia. If conditions were favorable for the growth of actinomycetes, it is plausible that fungi may also have grown in similar circumstances. (Some ongoing research uses similar principles to attempt to distinguish between wild and domesticated animals. Assuming domesticated animals were sometimes fed stored grasses and grains, the presence of tetracycline in excavated bone tissues may indicate domestication, and inversely, the absence of tetracycline could indicate that the animal was not domestic [H. P. Schwarcz, personal Web page, www.science.mcmaster.ca/geo/faculty/emeriti/schwarcz/research].)

The presence of fungi can plausibly be inferred from the presence of certain arthropods as well as from the presence of other microorganisms. That stored cereals and legumes were attacked by insect pests, both those from the field and those more specialized as storage pests, has been documented for Bronze Age Santorini, an island in the Aegean and site of a massive volcanic eruption that preserved by burial many artifacts (Panagiotakopulu and Buckland 1991). Other documentation is available for

Egypt and Roman Britain (Buckland 1981). Panagiotakopulu and Buckland (1991) make two points pertinent to the topic of fungi: that many of the insects identified from these ancient stores were likely the result of transport by humans along with seeds, and that extensive insect infestations promoted toxins from microbial infestation. Transport by humans in stored grain is also the most plausible reason for the appearance of certain insect pests in the archaeological record of the Roman period and preceding times, and damage by such insects is correlated with damage from toxic microflora (Buckland 1981). These are not just speculations; according to Christensen (1973), for example, "infestations by weevils and mites are almost inevitably accompanied by storage fungi." Indeed, even Theophrastus (1916b, trans. A. Hort) noted that rot of seeds was correlated with insect activity: "As seeds decay . . . they engender special creatures" and he mentions grubs, worms, etc. And Columella (1941a, trans. H. B. Ash) had much to say about mustiness and weevils in grain. Although insect pests have been documented in grain stores as far back as the Neolithic, these insects may not really have functioned as economic pests until grain was stored and/or transported on a large scale, perhaps by the Middle Bronze Age (Valamoti and Buckland 1995). Caution is essential in reviewing some aspects of the archaeoentomological literature, as some specimens identified as indigenous to a substratum, e.g. *Lasioderma serricone* in Egyptian tombs, are actually intrusive (Chaddick and Leek 1972, Kislev 1991).

Although most of the references cited here pertain to mycotoxins in grain or animal fodder, mycotoxins (especially ochratoxin A) are also important in wine. As noted in the sections above on fermented beverages, wine was a staple throughout much of the ancient world. Even in modern times, wines (especially red wines) are sometimes contaminated with ochratoxin A, especially when grapes have been grown in hot climates such as the Mediterranean region (Battilani and Pietri 2002). Species of *Aspergillus* and *Penicillium* are the fungi primarily responsible. It seems quite likely that similar or higher levels of contamination occurred in ancient times. Although many ancient wines were resinated, retsina wines do not appear to be resistant to production of ochratoxin A (Stefanaki et al. 2003). Moreover, although mycotoxins are most well known from their occurrence in cereal grain, they can also contaminate other stored solid foods. Flint-Hamilton (1999) speculates on the role of mycotoxins in lentil and bitter vetch in classical antiquity, citing some observations of Theophrastus in this regard.

Summary of Ancient Storage Practices

Early Neolithic grain storage was relatively primitive and consisted largely of sinking baskets or pots into the soil. Large reed baskets or clay jars were typical of the lower Nile delta in the Neolithic. Cylindrical earthen silos with openings in the roof were used in the First Dynasty (ca. 2920–2770 B.C.), and cylindrical chambers with vaulted roofs were in use by the Middle and New Kingdoms. Multichambered, flat-roofed storage facilities were also constructed ca. 2000 B.C., some with ramps leading to the roof (Nash 1985, Sinha 1995). Rickman (1971) writes of Egyptian methods, "There is every reason to believe that even in the structure and design of the granaries, Egyptian methods changed hardly at all during the centuries of Pharaonic, Ptolemaic and Roman rule." He lists two types, the circular ground plan (beehive-like) and the rectangular ground plan, both with provisions for adding grain at the top and removing it at the bottom.

Grain storage in Mesopotamia and the Levant ranged from pottery and pits to more elaborate rooms and silos. As early as the middle of the third millennium B.C., even some small settlements possessed fairly extensive storage facilities in the form of clusters of large subterranean or semisubterranean rectangular silos (Curvers and Schwartz 1990, Schwartz and Curvers 1992). Midsize settlements of Upper Mesopotamia stored grain in pottery, storage pits, and granaries (Wilkinson 1994). Rectangular, multiroomed brick and stone storehouses with granary bins characterized storage in a Middle Bronze Age site in Israel (Chernoff and Paley 1998). Chernoff and Paley (1998) repeat the oft-heard assertion that hulled grains (e.g., emmer wheat) were more resistant to postharvest losses than "naked" wheat. They discuss postharvest losses and state that insects and pathogens could be minimized in the granaries by sterilization of bins by fire (between storage cycles) or by proper deployment of ash, sand, or straw (the latter attested by phytoliths) to promote dryness. They nonetheless regard fungi as of reduced importance in the semi-arid climate of their site. They provide further examples of large-scale grain storage facilities in ancient Mesopotamia, Nubia, Macedonia, and elsewhere. Quantities could be considerable. Trade in grain between Mesopotamia and the Persian Gulf area in the third millennium B.C., though small by modern or even Roman standards, would suffice to sustain a few thousand people for several months (Edens 1992).

These archaeological examples are important for establishing storage practices in the Fertile Crescent, because although writing was well developed and transactions concerning grain were often recorded, textual references to grain storage are scarce: "Indeed, we hear surprisingly little about storage in early Mesopotamia in general" (Postgate 1992). Currid and Navon (1989), in addition to giving a brief synopsis of storage pits from ancient and modern cultures, provide details on storage pits from a variety of sites in Iron Age Palestine. They cull several references (usually indirect) to storage practices from the Old Testament. They also did experiments with pits excavated and constructed "like the usual Iron Age pit" (see "Experiments duplicating ancient storage practices" below). Iron Age Palestinians lit fires inside pits (between storage cycles, similar to the practice cited above) as a sanitary practice to eradicate pests.

The storage practices of the Minoan and Mycenaean civilizations were analogous to those of early Egypt. Very large clay jars (pithoi) were used to store food (and wine) in Neopalatial Crete (Christakis 1999). Although Linear A (the writing system of the Minoans) is yet largely undeciphered, ideographs denoting grain (and wine) appear to represent tallies of amounts of stored commodities (Nakassis and Pluta 2003). Stored commodities such as wine, olives, olive oil, grain, barley, wheat, and seed are denoted in Linear B, the early Greek writing system of the succeeding Mycenaean civilization (Palaima 2004a,b). Excavations of pre-Mycenaean structures at Lerna and Tiryns in the Peloponnese (i.e., the "House of Tiles" and other "corridor houses") and elsewhere, with fortified citadels and storage rooms, strongly imply an "organized trade structure" (Şahoğlu 2005) on mainland Greece. Wiencke (1989) notes that pithoi were excavated at Lerna, and she includes a concise description of the state of agriculture, indicating that in addition to cereals, the grapevine may have been domesticated there, or perhaps systematically exploited from the wild, as many seeds were recovered in excavations. This early system of production and trade apparently collapsed, but revived much later with the growth of Mycenaean culture.

Greco-Roman storage techniques varied from primitive to advanced. The Attic peasantry devoted a large part of their houses to food and seed storage, usually having their living space on the second (top) floor and storage on the ground floor. Large jars (pithoi) seem to have been used (Forbes and Foxhall 1995). The extensive use of hulled grain (e.g., emmer, spelt, hulled barley) in Greco-Roman antiquity may have helped to protect stored seed and food from insects and other pests but did not prevent deterioration. The Roman writer Columella (AD 60) advised against locating a villa close to a marsh, because the damp air could spoil stored produce (Forbes and Foxhall 1995). Such observations were relatively common in antiquity; e.g., Theophrastus (1990b, trans. Einarson and Link) advised, "The country is much better for keeping seeds if it is dry and cool." Pithoi sunk into the ground were more subject to damp and best avoided for grain storage, but they were sometimes used for that purpose (Cahill 2002). The Roman Varro (37 B.C.) commented that in Thrace and Cappadocia (Turkey), caves were sometimes used for grain storage, and in Tunisia and Greece, wells might be used (Nash 1985). Rickman (1971) has Varro describing such pits in Spain as well. According to Columella, pits could be used in dry climates but not in his country, Italy, because of the damp. However, Nash (1985) notes the extensive use of pits in Iron Age western Europe, including the roughly meter-deep "beehive" pits of Iron Age England. Wood (2000) reviews the use of pits in Celtic Europe and describes their basic construction and variations in storage conditions and efficiency; she also gives the highly optimistic report that the results of certain simulation experiments compared favorably with storage under the most modern conditions.

Although pits and other less sophisticated storage methods probably continued in wide use by the peasantry, Hellenistic and Roman grain storage facilities evolved into buildings (*horrea*) of considerable complexity and sophistication (for details, see Nash 1985, Patrich n.d., and Rickman 1971). Great attention was paid to ventilation (including raised floors and ventilators set in walls) and to the exclusion of rain and other sources of moisture. Husselman (1952) notes the dangers of spoilage, provides details on ventilation in terms of size of vaults, location of windows, and so on, and presents a detailed floor plan and cross section of the Roman-era granary at Karanis in Egypt.[5]

The precautions taken in construction demonstrate that the Romans were well aware of the potential for spoilage. According to Nash (1985), "Romans were frequently faced with the occurrence of mouldy grain in their large storehouses"; however, "they were probably unaware of the dangers associated with eating mouldy grain containing mycotoxins." Nash writes that storemen during late Roman times were instructed to mix bad, old grain with new grain, presumably to disguise as much as possible the off flavor of molded stores.

Grain might be stored for considerable periods of time. Forbes and Foxhall (1995) relate the comment of Theophrastus (*Inquiry into Plants*) that wheat seed kept for three years lost fertility but was still edible, and Nash (1985) relates Varro's statements that grain could be stored for up to 50 years and millet 100 years (!) under satisfactory storage conditions. Nash (1985) summarizes instances from Leviticus (2:14), Joshua (5:11), Roman Britain, and elsewhere of "parching" moist grain and notes the utility of parching before either storage or grinding. Unfortunately, precisely how grains were packaged in transport or storage is not well known. Rickman (1971) reviews what is known about storage in sacks, bins, or in loose heaps on the floor.

Experiments Duplicating Ancient Storage Practices or Using Ancient Crops

Hill et al. (1983) constructed underground, Iron Age-type storage pits in Hampshire, England, of various types (large and small cylindrical; large and small beehive; walls lined or not lined with clay, chalk, or basket

---

[5] Rickman (1971) gives some details that provide an appreciation of the scale of grain transport and storage: "by the end of the second century A.D. the standard size of ship used for the transport of grain had to have a capacity of at least . . . 340 [to] 400 tons." Some larger ships could handle 1,300 tons, and about 1,400 grain ships might dock at Ostia, Rome's port, each year, mostly from Egypt, Sicily, and northern Africa. In his comprehensive summary of grain trade in the Hellenistic era, Rostovtzeff (1941) says, "Hundreds if not thousands of ships were engaged in carrying corn [i.e., wheat and barley] from one port of the Hellenistic world to another." Rostovtzeff (1957) provides numerous details on and illustrations of transport and storage of grain in imperial Rome, e.g., a floor plan of a villa showing where grain was stored in large jars (dolia) and a mural showing grain being loaded onto a ship.

work, etc.). The pits were filled with barley grain, and grain was stored from October until April. Depending on pit type and location, temperature and grain water content varied considerably, as did the species of fungi recovered. *Penicillium verrucosum* var. *cyclopium, P. roquefortii, P. hordei, P. piceum,* species of *Aspergillus* of the *A. glaucus* group, and *Cladosporium* tended to increase in storage. Other taxa, mostly "field fungi" like *Aureobasidium pullulans, Alternaria* spp., and *Epicoccum purpurascens,* were isolated less frequently after storage than before, a result in accordance with more recent findings (e.g., Wicklow 1995). Pits within houses functioned better than pits outdoors, which were unsatisfactory in wet winters. In analogous experiments in Wiltshire, England, Lacey (1972) recovered *Aspergillus flavus, A. fumigatus, A. terreus, A. versicolor, P. cyclopium, P. decumbens, P. lanosum, P. piceum, P. rugulosum, Cladosporium* species, and several other fungi from barley grain stored in Iron Age-type pits dug into chalk. Unlike Hill et al. (1983), Lacey did not isolate fungi of the *A. glaucus* group. Grain was often moldy toward the edges of the pits, some of which were lined with basketry or straw. Lacey also summarizes some earlier literature on the topic.

Enzyme-linked immunosorbent assay (ELISA) has been used to detect aflatoxin B_1, ochratoxin A, and T-2 toxin on barley stored in Iron Age–type pits for six months (Brenton 1990). Wood (2000) reviews some literature on experiments with pits and reaches conclusions more optimistic than the above authors on their potential efficiency. With construction as described by Wood, grain "would keep for years without deteriorating," but under other conditions, grain germinates and "goes mouldy." Wood also describes fogous and souterrains, artificial caves used to store various foods or fermented beverages such as wines or meads in prehistoric Europe, and notes that cheeses stored in caves such as those at Roquefort were exposed to molds "that live in the caves of that region" and that imparted their unique flavor, presumably derived from *P. roquefortii,* to the cheese. Such cheeses were esteemed by the Romans (Kavaler 1965).

It should be noted in passing that the British in particular have long been fascinated by the possible purposes of ancient excavations. Well before mycotoxins were scientifically documented, Hayes (1909) reviewed the spectrum of opinion on certain excavations by the ancient Britons, and in so doing cited Tacitus on the use of storage pits by the Germans and several authors on ancient or modern storage pits in northern Africa, Syria, and Persia. Hayes recounted several early observations or experiments using pits to store grain, including the conditions under which storage was successful and those that led to decomposition. (The formation of an impermeable, hardened layer, or *croûte,* atop the grain was one precondition for successful storage. Wood [2000] mentions analogous sealing crusts.) Bersu (1940) was among the first modern scientists to present highly detailed excavations on Iron Age pits and to construct detailed and plausible explanations of their use in storing grains, nuts, and other foods in relation to preservation or decay. Speculation about ancient excavations is not at an end, as shown by the experiments of Marshall (1999) with grain storage. In this case the pits were large, ancient "silos" of a sort detected as magnetic anomalies by fluxgate gradiometry. Marshall briefly addresses aerobic versus anaerobic conditions in pits and the conditions necessary to avoid rotting of the grain.

Currid and Navon (1989), replicating conditions in Iron Age Palestine, experimented with varied pit designs (cylindrical or bell-shaped), linings (ash, stone, unlined), and storage durations (three months and up) and recorded the results in terms of temperature, moisture, and invasion by insects, rodents, and fungi. Although fungi were not identified, the percentage of grain infested by fungi over three months was recorded at 1.9–3.1% (mean 2.5%), a result regarded as minimal.

Other research has focused on mycotoxins in crops used by early civilizations. Spelt (*Triticum spelta*), emmer (*T. dicoccum*), and einkorn (*T. monococcum*) were among the earliest domesticated wheat, and their use persisted until the late Roman Empire, after which they were less

FIGURE 3

Penicillium verrucosum, a mycotoxigenic fungus. This organism is a common contaminant of stored cereal grains. (From Frisvad and Samson 2004; reprinted by permission of the authors.)

widely planted. These three plus wheat, rye, barley, oats, and triticale have been analyzed for ochratoxin A. Elmholt (2004) tested samples from market grain destined for specialty bread production in a Danish bakery. The percentage of kernels or spikelets colonized by *Penicillium verrucosum* (Figure 3) was recorded on the basis of symptoms and recovery of the fungus onto agar medium, while high-performance liquid chromatography (HPLC) was used to analyze for ochratoxin. Spelt was contaminated with *P. verrucosum* more than other grains, in spite of the retention of glumes, paleas, and lemmas on a high proportion of threshed kernels. There was no linear relation between rate of contamination and level of ochratoxin, although *P. verrucosum* was always detected in samples producing ochratoxin. Elmholt concludes that "covered wheat" (i.e., retaining glumes, paleas, and lemmas) does not derive substantial protection from fungi compared to grains for which threshing produces naked kernels. Elmholt and Rasmussen (2005) also found spelt significantly more contaminated by *P. verrucosum* than other wheat, oats, or barley. In similar fashion, Castoria et al. (2005) researched the occurrence of fungi and mycotoxins in stored farro (a collective name for *T. monococcum, T. dicoccum,* and *T. spelta*) from southern Italy. Potentially mycotoxigenic species *Fusarium proliferatum, F. tricinctum, P. verrucosum,* and *P. chrysogenum* were recovered, as were *Aspergillus niger* and *A. tamari.* Ochratoxin A (produced by the *Penicillium* and *Aspergillus* species) and fumonisin B_1 (produced by *F. proliferatum*) were detected, as well as some other mycotoxins.

Ancient Historical Events Speculatively Attributed to Mycotoxins

Bellemore et al. (1994) give a detailed and speculative account of one spectacular and, if true, historically significant episode of mycotoxicosis. They analyzed Thucydides's *History of the Peloponnesian War* in detail with regard to the timing, symptoms, and circumstances of the great plague of Athens during the Peloponnesian War. They conclude that the plague was not, in fact, a contagion but was plausibly induced by *Fusarium*-molded wheat, with symptoms of mycotoxicosis closely matching those of alimentary toxic aleukia (ATA). Arguments in favor of the hypothesis include that because Athens was under siege, the primary foodstuff for the upper classes (who were very disproportionately affected) was wheat imported from the Black Sea area. The wheat was described by another historian, Diodorus Siculus (12. 58. 3–5) as bad. That the besiegers did not contract the plague is additional evidence. ATA is known as endemic to grain-producing areas of Russia and Ukraine to the present day. Grain from southern Russia was widely imported throughout the Mediterranean in Hellenistic times and later (Grant 1970, Rostovtzeff 1941). Schoental (1994) also points to the plausibility of mycotoxins with regard to the plague at Athens. Salway and Dell (1955) proposed that ergotism was the cause of the plague. It should be noted, however, that literature speculating on the plague of Athens is abundant, and there are indeed references to the plague acting as a contagion (e.g., Thucydides himself attempted to explain how the plague traveled to Athens, and soldiers at the siege of Potidaea were said to have contracted plague from Hagnon's expeditionary force). But most sources have noted that it was not highly contagious; e.g., Cruse (2004) describes it as "contained" and "local."

In another highly speculative approach, Schoental (1984) gives some evidence from the Book of Job in favor of an etiology involving mycotoxins in the sufferings of Job. Schoental (1980) had previously argued that Mosaic dietary prohibitions "appear now as if . . . designed for protection from mycotoxins." Schoental (1991) also proposes a role for mycotoxins in the decline of the Etruscans. The hypothesis involves toxic metals as well as toxic fungal metabolites, with an eventual effect on the sexual organs and sexual orientation of the Etruscans, resulting in decline vis-à-vis the increasingly powerful Romans. This latter hypothesis has been d'scounted by Spivey (1996). Yiannikouris and Jouany (2002) briefly note the above accounts on Athens and the Etruscans in their review of mycotoxins in feeds.

Schoental (1987) also considers the impact of mycotoxins on the lifespan of individuals in the Bible, arguing that mycotoxigenic fungi flourished in the moist environments after Noah's flood. Before this event, individuals such as Adam, Methuselah, and Noah had ample lifespans of greater than 900 years, but after the flood persons went to their graves after a trifling 150 to 400 years or so. Schoental carefully documents the lifespan statistics with verses from Genesis. This particular hypothesis from Schoental is creditable only in the context of Biblical literalism.

Another attempt to explain Biblical events by reference to mycotoxins is that of Marr and Malloy (1996), who postulate that exposure to *Stachybotrys atra* (a strongly toxigenic fungus) might provide a rational explanation for the tenth plague of Egypt (death of the firstborn) as recounted in Exodus. Essentially, the argument is that the Egyptians responded to the prior disasters by hoarding damp grain, which became contaminated with mycotoxins. The eldest children, receiving preferential treatment in such times, would have received the most food and thus the highest dose of trichothecene mycotoxins. Schoental (1980) also mentions the possible relevance of mycotoxins to the Biblical plagues of Exodus and the Pentateuch.

It is difficult to find direct evidence for impacts of T-toxins (trichothecenes), fumonisins, or ochratoxin on ancient peoples, although the above review makes it plausible that negative effects sometimes occurred. There is, however, some evidence for effects from toxins derived from *Claviceps* species, especially *C. purpurea*, cause of ergot. (See also Fungi as "Entheogens.") Brothwell and Brothwell (1969) note that the better-documented ravages of ergotism in the Middle Ages strongly imply analogous negative impacts in antiquity. They mention (without attribution) Babylonian and Assyrian tablets alluding to "noxious grasses" and "noxious pustule in the ear of grain," respectively. The allusion to "noxious pustule" has also found its way into the literature on mycotoxins (Bennett and Klich [2003], citing Hofmann [1972], who also wrote of the "noxious pustule" without attribution). Brothwell and Brothwell also point out probable allusions to ergot in Theophrastus, Hippocrates, Pliny, and Galen, but state that the first clear account of ergot was given by the Perso-Arabic physician Muwaffak (AD 950). Carefoot and Sprott (1967) disagree, writing that ergotism was probably absent from ancient Mediterranean civilizations before invading barbarians (Franks, Vandals) brought rye (and with it, ergot) during their incursions into Roman territory. Matossian's (1989) masterful study of ergotism does not much explore times prior to the Middle Ages, but she states (without attribution) that in Roman Gaul the ruling classes preferred wheat to rye and that even in that era, people were aware that ingestion of rye could be hazardous. Gangrene of the extremities is among the possible symptoms of ergot poisoning, and "Galen spoke of coloured grain causing something like gangrene" (Aaronson 1989). Verster (1976) argues that miscellaneous descriptions of psychosomatic disturbances in classical mythology and literature are in accord with symptoms of ergotism.

Ergot and ergotism are most closely associated with rye (Matossian 1989), and although rye is attested early in the archaeological record in the ancient Near East and eastern Europe (Zohary and Hopf 2000), its cultivation was not widespread in antiquity, except perhaps in some northerly regions. However, weedy forms were prevalent and admixed with other cereal crops (Zohary and Hopf 2000). Zohary and Hopf also report convincing excavations of rye from Neolithic sites in Poland and Romania and from Bronze Age sites in the Czech Republic and Slovakia. Aaronson (1989) compiled records of Neolithic, Bronze Age, and Iron Age sites from which *C. purpurea* sclerotia were recovered in excavations in Poland, Scotland, Germany, Sweden, and Denmark. The fungus was found in association with rye, wheat, barley, and wild grasses. Dark and Gent (2001) cite further instances of ergot sclerotia recovered from ancient sites in the Netherlands and elsewhere and note that sclerotia were found in the stomach of one of the "bog bodies" from the Danish Iron Age (citing Helbaek 1958). The recording of ergot sclerotia in the digestive tracts of bog bodies has produced an apparent controversy over whether such persons were drugged before their demise (Fischer 1987).

Wheezing and Sneezing: Molded Feed, Hay, and Silage

Pits were probably used to produce silage as well as to store grain. The cumulative evidence renders early production of silage or stored hay plausible. Silage may have been produced in Egypt around 1500–1000 B.C., as depicted in an ancient mural, and in ancient Carthage and Crete at similar dates, as attested by remains of silos (Schukking 1976). Isaiah 30:24 has been translated to imply use of silage: oxen and donkeys are said to eat "provender," which is variously described as savory, salted, clean, or fermented, depending on the translation.[6] Apparently it is still debated whether silage was produced in early Neolithic Britain (Matthews 1997). Remnant dairy fats on Neolithic pot shards from central and eastern Europe are evidence for consumption of dairy foods and retention of livestock as dairy producers (Craig et al. 2005), making some form of winter feeding of livestock plausible, i.e. silage or other stored fodder. Storage pits may have been used in the Iron Age to produce silage in Britain (Reynolds 1976). By Roman times there were numerous writings on the proper harvest and storage of fodder, and Cato noted that the Germanic peoples would store green fodder in pits dug into the ground and covered with dung (Schukking 1976). Although there is little direct documentation for silage production between A.D. 100 and the Renaissance, it is likely that ensiling of forage crops continued "on a small scale" (Wilkinson et al. 2003).

Several species of fungi are capable of producing mycotoxins in hay and improperly stored or manufactured silage (Pahlow et al. 2003, Scudamore and Livesey 1998). In addition to mycotoxicosis (including mycotic abortion in cows), livestock, especially horses, are subject to other diseases, including chronic obstructive pulmonary disease or pulmonary emphysema (commonly called "heaves" in livestock), an allergic response to heavily molded hay (Ward and Couëtil 2005). Consumption of moldy grain or hay is also the most likely source of zygomycosis in animals (Ribes et al. 2000).

Ainsworth (1986) chronicles the history of diseases in humans caused by allergies to fungi, including "farmer's lung" or "thresher's lung" and allergies from fungi on grain stored in pits. The ancients were, of course, unaware of the microbial causes of allergenic asthma but, as documented below, they were aware of the symptoms in humans and animals. And, of course, they were also aware that moldy substances were highly distasteful. When Josephus described the conditions of deprivation during the Roman siege of Jerusalem, he mentioned that some persons were so driven by hunger that they consumed moldy hay, a substance that "even . . . dumb animals were not wont to eat" (Pike 1938).

There are references to asthma in Hippocrates, Pliny, and Galen, but Aretaeus of Alexandria gave the first tolerably accurate description of "the constellation of symptoms comprising the asthma syndrome" (Sampson and Holgate 1997). Marketos and Ballas (1982) discuss what

---

[6] The use of "fermented" is from Young's Literal Translation of 1898, but the King James Version merely says "clean provender." Zohary (1982) argues on philological grounds that the reference to "provender" is mistaken and that the original Hebrew referred to chickpea (*Cicer arietinum*). Moldenke and Moldenke (1952) also discuss putative meanings of this passage.

is known about asthma in writings of Aretaeus and other physicians of Greek antiquity, but Berstein (2003) emphasizes that it was not until the Italian Renaissance that medicine recognized the connection between certain occupational environments (grain threshing, handling old clothes, etc.) and pulmonary diseases. However, the veterinarians of Greece and Rome described symptoms of pulmonary emphysema (heaves) in horses and recommended avoidance of spoiled hay in treatment of colic (Moule 1990), although other approaches to animal health at that time were likely ineffective (bleeding of the animal, treatment with cat blood, etc.).

It cannot be assumed that peoples in ancient times exposed to fungal spores originating in contaminated grains, silage, fodder, and the like reacted to them exactly as modern humans do. The evolution of allergenic response with changing living conditions (from the Neolithic to modern times) is controversial, and a number of hypotheses have evolved that integrate changes in diet, exposure to parasites, air pollution, and other factors (Armelagos and Barnes 1999).

Fungi Used for Medicinal Purposes and Other Technologies

Well before the age of antibiotics, ancient peoples were able to appreciate the medicinal properties of some fungi. However, in the absence of controlled experiments, there was a strong tendency to regard any medicinal plant or fungus as a panacea for a large variety of ailments. The *agaricum* of the Romans was thus credited with the power to heal everything from broken bones to mental illness to flatulence. Other fungi, such as some boletes, had the same "medicine show" appeal. In fact, the Romans seem to have anticipated the patent medicines of the late nineteenth and early twentieth centuries by producing mixtures of ethanol (mostly wines of various sorts) with other drugs (principally opium), and these medicated wines were an item of extended trade.

Potentially more rational, since we now understand that many molds are producers of powerful antibiotics, was the application of moldy bread or other materials to surface wounds to combat infection. In practice, however, placing any rotting matter onto a wound probably had as many hazards as benefits. The use of the tinder fungus to cauterize wounds may also have been beneficial in isolated cases, but in other instances was probably just a form of medically sanctioned torture. The more prosaic uses of the tinder fungus to start fires and produce felts no doubt had more reliable benefits. Interestingly, the use of conks for starting fires seems to have very ancient roots, going back even to the Paleolithic, with continuity to modern times. Tracing these latter practices is highly revealing in matters of mythology and folklore. There are abundant examples in European folklore of the use of fungi for medicinal purposes, but the degree of antiquity of many of these practices is highly uncertain. Ancient peoples also used some lichens as medicines and fabric dyes.

Documentation: Fungi Used for Medicinal Purposes and Other Technologies

Panaceas and Pyrotechnics

Most reviews of fungi and ancient medicine focus primarily on the *agaricum* (*agarikon* in Greek) of the Romans. *Agaricum* was almost certainly *Laricifomes officinalis* (syns. *Fomitopsis officinalis, Polyporus officinalis*) (Figure 4). The first surviving documentation of *agaricum* is by Dioscorides (Houghton 1885, Phillips 1982). Buller (1915), Houghton (1885), Pegler (2000b), and Phillips (1982) provide synoptic translations of the pertinent portions of Dioscorides's writings, including the conditions against which *agaricum* was effective (colic, sores, bruises, fractured limbs, stomachache, dysentery, asthma, hysteria, complexion problems, flatulence, and others), its dosage (with or without wine, honey, or other liquids), and its usefulness as an antidote to poisons. The opinions of Dioscorides were perpetuated in Galen and Pliny. The *Agaricus* [*sic*] mentioned by Ramoutsaki et al. (2002) in their review of remedies for otolaryngological

FIGURE 4

Laricifomes officinalis, the *agaricum* of the Romans. The fungus is distributed throughout the Northern Hemisphere; this specimen is from British Columbia. (Eric Allen ©1996 Her Majesty the Queen in right of Canada, Natural Resources Canada, Canadian Forest Service)

problems of the Byzantine period is probably *agaricum*. The interesting but erroneous statements by Dioscorides on the sexes of *agaricum* were analyzed by Buller (1915), Houghton (1885), and Pegler (2000b).

Pegler (2000b) documents analogous medicinal uses of this fungus by American Indians of the Pacific Northwest, as well as the spurious trade in "PseudoAgaricum" (*Laetiporus sulphureus*) as a dishonest substitute during the early Renaissance. *Laricifomes officinalis* was part of the pharmacopoeia of the nineteenth and early twentieth centuries (Buller 1915) and was used as late as World War II as a source of quinine against malaria (Blatner 2000).

Other fungi were also credited with medicinal value in Greco-Roman tradition. Pliny recommended *suilli* (probably *Boletus edulis* or related fungi) for problems of the digestive tract, blemishes, sore eyes, and other complaints (Buller 1915). Molds growing on bread or grains were used to treat surface infections or wounds, including instances from ancient Egypt (from the healer Imhokep) and in the Talmud, and such practices persisted into later eras (Wainwright 1989a). Similar medicinal applications of molds occurred in ancient Greece and figure in English folklore (Wainwright 1989b). Aaronson (1989) cites several instances of the use of smut fungi for medicinal purposes in southern or eastern Asia and acknowledges the potential use of ergot in some instances but does not document specific uses in west Asia, Europe, or the Mediterranean. There are numerous uses of fungi in folk medicine, although it is unclear how ancient most of these practices were (Morgan 1995).

Fungi had other functions in ancient medicine. Buller (1915) was of the opinion that Pliny's reference to the use of fungi for starting fires was to *Fomes fomentarius* (Figure 5). This fungus has been used as tinder up until modern times. Moreover, Buller thought it likely that Hippocrates

FIGURE 5

Fomes fomentarius, the tinder fungus. This species was used from at least Mesolithic times for kindling fires. (From Nees von Esenbeck 1816–1817)

had this same fungus in mind when recommending cauterization for specified ills, including those of the liver and spleen. In fact, cauterization with *F. fomentarius* was recommended for a number of maladies and apparently constituted the first documented use of fungi in ancient medicine (Phillips 1982). During the Neolithic and into near-modern times, puffballs (*Bovista* spp.) may have been used in analogous fashion for making or preserving fire and for medicinal purposes; the Lapps also used *F. fomentarius* for cauterization (Morgan 1995). *F. fomentarius* was also one of the fungi found among the equipment of the "Iceman" preserved in glacial ice some 5,000 years ago (Peintner et al. 1998). It had been transported in a small leather bag worn by the Iceman. Other fungi, found as fruiting bodies strung on leather thongs, were *Piptoporus betulinus* (Figure 6). This fungus, edible when very young, also had medicinal uses in folk medicine of much later times, but its intended use by the Iceman is speculative. It is possible that the fruiting bodies had both medicinal uses and spiritual connotations (Peintner et al. 1998).

Graves (1960) discusses a scene from an Etruscan mirror that portrays Ixion bound to a fire-wheel "with mushroom tinder at his feet." (In classical mythology, Ixion was punished for attempting the seduction of Hera.) Graves recounts parallel fire-wheels in European folklore

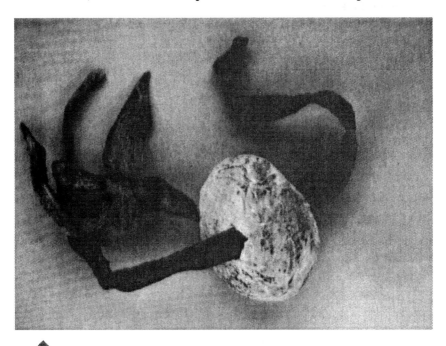

FIGURE 6

Piptoporus betulinus, on a leather thong with tassel, from the Iceman's apparel (ca. 3000 B.C.). (From Peintner et al. 1998; reprinted by permission of the British Mycological Society.)

and alludes to connections with ritual sacrifice. Later, expounding on tales regarding Sisyphus, he notes that Sisyphus and Ixion were placed together in Tartarus and uses this to invoke a connection between these two mythological figures, mushroom tinder and fire-wheels, and the "birth" of the people of Ephyra (Corinth) from mushrooms. Houghton (1885) linked the production of bodies from mushrooms in Corinth with a different mushroom, *Phallus impudicus*, because of "the licentious nature of the people and the extent to which the worship of Aphrodite prevailed in the city."

The use of polypores for tinder or other purposes probably has an even more ancient history. *F. fomentarius*, *Phellinus igniarius*, and other polypores were recovered from Neolithic as well as Bronze Age contexts in France. Apparently the presence of the fungi at these sites was not incidental, and production of fire is considered their plausible use (Monthoux and Lundström-Baudais 1979). Bernicchia et al. (2006) identified *Daedaleopsis tricolor* from remains at a Neolithic village in Italy and speculate on its possible pharmacological use by the inhabitants. And Corner (1950) documented *F. fomentarius* "in quantity" from the Mesolithic Star Carr site (U.K.): "It might therefore be that the specimens . . . had been collected . . . for some purpose, and it is significant that some of the specimens have the . . . upper layer of the bracket . . . stripped off from the tubes, as mentioned in Ramsbottom's account of *amadou*." Amadou is a feltlike product manufactured from this fungus. Roussel et al. (2002) review the history of usage of *F. fomentarius*, including Roman, Greek, and later contexts up to modern times. Bracket fungi even found a use in diplomacy. Cassius Dio (*Roman History*, 68.8.1) told of an incident in which barbarians wrote a message on a fungus to the emperor Trajan, urging him to desist in his attacks. It is only conjecture, but the mushroom may have the "artist's conk," *Ganoderma applanatum*, whose lower surface turns dark brown if inscribed.

Use of "mushrooms" as tinder persisted in western Ukrainian rituals (Rubchak 1981a, n. 19). In a letter to Maxim Gorky, Mykhailo Kotsiubynsky wrote, "The Hutsul [a Ukrainian tribesman] is a profound pagan; he spends all his life battling evil spirits that dwell in forests, mountains and waters. He uses Christianity only to decorate his pagan cult" (Rubchak 1981b). Rubchak (1981a) describes the fire-starting ritual in detail. It is highly likely that this practice occurred in eastern Europe well back into pagan antiquity. A similar instance at the opposite end of Europe involved "a species of agaric [*sic*] which grows on old birch trees, and is very combustible" that was used to initiate the Beltane fires of the Scottish Highlands (Frazer 1955e). The Beltane fires were a "curious and interesting picture of ancient heathendom surviving in our own country" and may be related to the ritual burning of toadstools (*Bären*) in analogous fire festivals in Sweden (Frazer 1955e). Toadstools were used in this manner to counteract the power of trolls. Morgan (1995) provides details on the probable species of fungus (*Tremella mesenterica*) and specific attributes of the trolls. (For more on *Tremella*, see Fungi in Ancient European Folklore.)

It is possible that the fungus *Inonotus obliquus,* the "chaga" marketed on many Web sites, was used in Russia in very ancient times. Vaidya and Rabba (1993) discuss the fungus in a context implying that it was used since antiquity ("since ancient times dating back to the Greeks and Romans"), but Pegler (2001) writes only that "chaga has been used for medicinal purposes from at least the sixteenth century." It is often difficult to trace the origins of folk medicines. Chaga is held to be effective in relieving pain and sickness in people with tumors (Pegler 2001).

Perhaps not surprisingly, fermented beverages, especially wine, were also used medicinally. Medicated or drugged wines probably have an ancient history. Helen, returned from Troy and again the queen with Menelaus at Sparta, served a medicated wine to Telemachus to temporarily relieve his grief over the absence of his father, Odysseus. Wine was amended with opium (and perhaps *Cannabis*) in Roman times (Cruse 2004). "Aminean wine"—wine amended with horehound—was considered a cure-all but especially good for diarrhea and the common cold. Cato (234–149 B.C.) gave several recipes for amending wine with black hellebore (probably *Helleborus niger*) for use as a laxative. *H. niger* possesses drastic purgative properties; it was either applied to vine roots or added directly to the must. Cato (1934, trans. W. D. Hooper) assures us that if we follow his recipes, the wines "will move the bowels with no bad results." Medicated wines were widely traded throughout the Roman world (Cruse 2004). A wine and vinegar mix was used as an eye salve. Other examples mentioned by Cruse (2004) include the Egyptian application of "fermenting yeast" as part of a poultice for treating burns, and onion and honey "taken in beer" for local inflammation.

Lichens as Remedies and Fabric Dyes

There are records of fungi, e.g. lichens, used to dye fabric in ancient times. A prime example is the lichen *Ochrolechia tartarea* ("cudbear"), used in Celtic countries and Scandinavia (Ferreira et al. 2004) as a purple dye (giving "false shellfish purples"). It was known to the Romans and has been tentatively identified on fabric from a "bog body" from the Roman era (Turner et al. 1991). There are also Biblical and classical (Theophrastus, Pliny) references to lichens attributable to the family Roccellaceae being used as dyes (Sharnoff n.d.).

In addition, lichens had medicinal and funerary uses. Hippocrates recommended the lichen *Usnea barbata* for uterine trouble. Egyptians used *Pseudevernia furfuracea* for packing into the body cavities of mummies, to preserve the odor of spices in such embalming, and they even seem to have sometimes employed this same lichen when making bread (Sharnoff n.d.), a practice continuing into modern times (Darby et al. 1977). The lichen *Cetraria islandica* was used in ancient times as a cough remedy in Europe (Fern n.d.).

Plant-Pathogenic Fungi

Fungi as agents of plant disease have undoubtedly had a significant impact on human affairs since the beginnings of the Neolithic, and that impact has risen along with the reliance of most societies on plant agriculture. Although the most famous examples are relatively recent events such as the potato blight in Ireland, numerous examples from ancient writings and the archaeological record show that fungal plant pathogens had a strong effect on ancient peoples. It is both heartening and heart-rending to read the accounts of ancient authors, to appreciate their sometimes considerable powers of observation, but also their powerlessness in the face of unseen or misunderstood forces.

The coevolution of fungal plant pathogens and agricultural plants has not only been central to the historical development of agriculture, but also exerts a controlling effect over large areas of modern plant breeding and plant pathology. Plant breeders and plant pathologists continue to explore the countries of the Fertile Crescent for plant genes and races of pathogens. Scientists who explore this region, either by searching for landraces in situ or by proxy via obtaining such plant germplasm from seed banks, are essentially mining the distant past in order to provide benefits for the future. Some modern scientists have even advanced claims that spatial and temporal locations within the Neolithic can be determined for the origin of certain *formae speciales* (strains within a given fungal species that are specialized to cause disease on a given plant host species), races (analogous strains specialized on a given cultivar of the host species), or resistance genes in the host.

Just as dramatic claims have been advanced for the influence of fungi on the development of ancient religion and culture, equally strong claims have been made for the impact of fungal plant pathogens on ancient cropping systems. Because there are repeated references to plant disease in the Bible (and in other ancient writing), much ink has been expended in attempts to assign specific causes to these descriptions. Scholars have not always agreed on the extent to which success is possible, and the most skeptical have stated that most such efforts will never yield definitive conclusions. In a few rare instances, archaeologists have been able to recover identifiable spores of plant pathogens from ancient substrata. Fungi belonging to genera known as facultative plant pathogens have also been recovered from ancient glacial ice or permafrost (see Ancient Fungi Preserved in Glacial Ice or Permafrost).

Documentation: Plant-Pathogenic Fungi

During and since the Neolithic, most peoples have derived the bulk of their caloric intake from crop plants, and so they early on became acute observers of plant diseases. Although limitations of early technology precluded knowledge of the specific microbial causes of plant diseases, the ancients documented symptoms and integrated observations into their existing beliefs and practices. Harlan (1976) characterizes the changes in land use during the Neolithic as a "manmade imbalance" that imposed an artificial flora over extensive areas and thereby created abundant opportunities for extensive reproduction of plant pathogens. These opportunities grew apace as villages coalesced into towns and fields became more extensive. Modern plant pathologists and geneticists have sometimes been able to use the tools of molecular genetics to trace the evolution of specific pathogens during Neolithic times.

Ancient Agriculture, Landraces of Crop Plants, and Fungal Plant Pathogens

As noted in the Introduction, modern scientists have assumed that fungal pathogens (or other pests) and their hosts have coevolved, and they have taken a strong interest in the geographic centers of origin of foundation crops and allied species in their search for plant cultivars resistant to certain fungal pathogens. This is especially true for fungal pathogens (or symbionts) of crops in areas of the Fertile Crescent and the Mediterranean, such as wild barley species or barley landraces (Bonman et al. 2005; Clement et al. 2004; Czembor 2001, 2002; Fetch et al. 2003) or wild *Triticum* and landraces of wheat (Bonman et al. 2006, Lamari et al. 2003, Marshall et al. 1999). Harlan (1976) gives a very readable synopsis of the utility of ancient crop plant cultivars (landraces) with respect to breeding for resistance to fungal diseases of plants. Harlan himself collected a wheat line, PI 178383, with resistance to some races of stripe rust, common bunt, and dwarf bunt pathogens, plus tolerance to snow molds. (Readers can find many analogous examples on the Web site of the USDA-ARS National Plant

Germplasm System by searching for "landrace" plus descriptors for host, pathogen, geographic location, etc., at www.ars-grin.gov/npgs.)

Occasionally, researchers venture to speculate on an ancient spatio-temporal origin of a given gene; for example, Chartrain et al. (2005) suggested that the *Stb6* wheat resistance gene to *Mycosphaerella graminicola* arose in southwest Asia in the mid-Neolithic. There is strong evidence that the center of origin of *M. graminicola* coincides with the center of origin of its host (Banke et al. 2004, Zhan et al. 2003). Jiménez-Gasco et al. (2004) present two highly specific alternative hypotheses for how the known races and pathotypes of *Fusarium oxysporum* f. sp. *ciceris* (a fungal pathogen of chickpea) originated in the Fertile Crescent (the center of origin of domesticated chickpea) and spread in the Neolithic to adjacent areas. Wyand and Brown (2003) place the divergence of different *formae speciales* (ff. spp.) of the powdery mildew fungus *Blumeria graminis* as taking place over a period of up to 14,000 years, i.e., beginning at the very earliest stages of the Neolithic. Interestingly, they conclude that strict coevolution was absent in the case of *B. graminis* and its hosts, because the *B. graminis* ff. spp. most closely related by molecular-genetic characters could occur on grass family hosts that were not correspondingly related phylogenetically. The possibility that domestication initiated in the Neolithic may have led to changes in host-pathogen interactions at the genetic level has also been asserted for the Americas (Gepts 2004).

It should be noted here that instances of putative coevolution of plants and fungal pathogens during the Neolithic are not reflective of fungal evolution in general, which probably proceeded much more slowly. For example, Scherrer et al. (2005) apply the term "recent" to events of one to two million years B.P. with regard to the rust resistance locus *Rph7* in barley. The formation of cryptic species (sometimes called emergent species, i.e., species that strongly resemble each other morphologically but do not interbreed) in *Amanita muscaria* probably dates from about 7.5 ± 4.5 million years ago (Geml et al. 2006). It should also be stressed that not all pathogens invariably originated in the same geographic areas as their current agronomic hosts. There is evidence, for example, that the center of origin of the barley scald pathogen, *Rhynchosporium secalis*, may be in central or northern Europe, whereas cultivated barley has its center of origin in west Asia (Zaffarano et al. 2006).

The Ancient Near East

Orlob (1973) provides the most comprehensive synopsis of the writings of ancient peoples on plant pathogens, including fungi, and his summaries also cover writings of medieval times, India, China, and pre-Columbian America. For example, Orlob selects this translation from writings of the Sumerians: "If the . . . barley has turned red . . . it is sick with the samana disease" (from Kramer 1963). The goddess Ninkilim was invoked for assistance. Other examples provided by Orlob include a possible reference to lodging in barley (again from Kramer 1963) and an incantation against "mehru" disease, translated by Landsberger and Jacobsen (1955) as ergot, but according to Orlob, "possibly ergot, smut, head blight, etc." Orlob also briefly summarizes references to fruit rot (from Kramer 1956) and apparently abiotic damage from sun and wind. From James (1962), Orlob notes an Egyptian Middle Kingdom reference to reddening of grain and apparent lodging. Other, more cryptic references to crop damage have been located in Egyptian writings, but the phrasing is so embedded in religious or mythological context that actual symptoms are not apparent.

The Bible makes numerous references to plant maladies (summarized in Moldenke and Moldenke 1952, Orlob 1973). The most cited are "seven thin ears came up blasted" (Genesis 41:23); "The Lord shall smite thee with . . . blasting, and with mildew" (Deuteronomy 28:22); "if there be blasting, mildew, locust . . ." (1 Kings 8:37); "If there be dearth in the land, if there be pestilence, if there be blasting or mildew, locusts or caterpillars . . ." (2 Chronicles 6:28); "I have smitten you with blasting and with mildew; I laid waste your gardens and your vineyards; and your fig and your olive trees, the locust devoured" (Amos 4:9); and "I smote you with blasting and with mildew" (Haggai 2:17). Orlob (1973) cites literature that attempts to identify causal organisms of plant diseases from Biblical texts, but he is skeptical of the results. Orlob also reviews references to plant diseases in the Talmud as collected by Goldschmidt (1897) or Goor (1967), including references to blighted grain, destruction by blight and mildew, putative damage to wheat and olive from winds of specified direction, and damage or disease on pomegranates. Carefoot and Sprott (1967) comment on some specifics of translation from the Hebrew *yeraqon* (yellowing) to the King James "blasting" or "mildew" and conclude that references to blasting or mildewing therefore most likely indicate rusts.

Ainsworth (1981) comments at length on the difficulty (sometimes impossibility) of interpreting early records of plant disease and emphasizes that "the inclination to strain hindsight, when not resisted, frequently leads to more precise interpretation than the evidence would appear to justify." He comments on the ambiguity of terms such as mildew, blight, and blasting and their confusion with more modern, more specific terms like rust, smut, powdery mildew, and downy mildew. Like Buller (1915) and Houghton (1885), Ainsworth drew his readers' attention to confusions introduced by Linnaeus and by common usage.

Although it is doubtful that specific causal agents can ever be conclusively identified solely from descriptions in ancient literature, there are a few archaeological records of plant-pathogenic fungi from the ancient Near East. *Puccinia graminis* f. sp. *tritici* was present on lemma fragments of wheat (*Triticum parvicoccum* [syn. *T. turgidum*]) from the late Bronze Age (ca. 1400–1200 B.C.) in Israel (Kislev 1982). Kislev (1982) places archaeological evidence in the context of references in the Bible and the Mishnah, as well as references from classical antiquity, and notes analogous archaeological finds of *Mycosphaerella*-infested *Lolium perenne* grass (ca. 1000 B.C.) and fungal hyphae in darnel grain (ca. 2000 B.C.) in Egypt.

Fungal spores have been recovered from ancient potsherds by the use of acetate film (Stewart and Robertson 1968). Although all traces of the protoplast had disappeared, several taxa were discerned in these shards from Jarmo, Ali Agha, Sarab, and Matarrah in Iraq. Recognizable were spores of *Alternaria, Cladosporium, Helminthosporium sativum,* and *Puccinia graminis,* the latter two being common (Figure 7). The probable source of the spores was plant material used for temper in pot manufacture. *H. sativum* (syns. *Drechslera sorokiniana, Bipolaris sorokiniana*) and *P. graminis* are parasites of barley and emmer wheat, which have been reported from the sites. Leonard and Szabo (2005), citing Chester (1946) on Biblical references and Kislev (1982) on archaeological evidence, treat it as fact that the Bible referred to rusts and smuts. Spores of *Tilletia caries* (agent of common bunt) and *Urocystis agropyri* (agent of flag smut) were found in tufa deposits from late Paleolithic Egyptian Nubia. Both fungi are pathogens of several species of wheat, and pollen and trichomes typical of wheat were also present in the tufa deposits (Purdy 1968). Bunt spores have been found on seeds dating from 4,000 years ago in ancient Mesopotamia (Borgen 2000, citing Johnsson 1990).

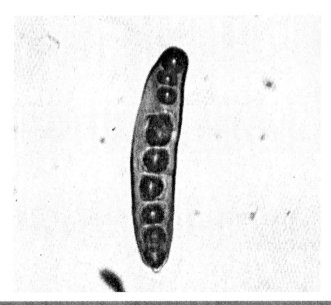

FIGURE 7

Helminthosporium sativum (Bipolaris sorokiniana) from a Neolithic pot shard (Jarmo, Iraq). (From Stewart and Robertson 1968; reprinted with permission from Mycologia; © The Mycological Society of America.)

Greco-Roman Antiquity

Little survives in classical Greek writings of relevance to plant pathology. Orlob (1973) cites references to grain that droops and the fall of green fruit, both in the context of the sorrows of Demeter, the normally benevolent goddess associated with crops and farming. Plato's *Symposium* makes reference to the ill effects of discord among the elements, such effects including diseases and mildew; Kleidemos (440 B.C.) had a humoral theory of plant health and believed that improper nutrition could lead to disease in fig, olive, and grape; and the so-called *Lithica* (ca. 400 B.C.) asserted that agate prevented hail and pests arising from the soil (Orlob 1973).

Beginning with Aristotle (384–323 B.C.) in the Hellenistic era, there are more direct references to plant disease, although Orlob points out that some scholars believe that Aristotle's writings on plants are actually pseudo-Aristotelian and originate later in the first century B.C. Citing Ross's (1913) treatment of *De Plantis,* Orlob notes belief in transmutation (including wheat into tares, probably *Lolium temulentum*) and miscellaneous references to blight and dry gangrene of plants.

Theophrastus (372–285 B.C.) was the first to classify fungi in a category distinct from plants, and his taxonomic treatment of fungi was not surpassed until that of Ray in the seventeenth century (Morton 1981), although some scholars believe that parts of the works of Theophrastus were written by Aristotle. Both of Theophrastus's main works, *Historia Plantarum* (often called *Inquiry into Plants*) and *De Causis Plantarum,* frequently refer to plant disease. For example, Theophrastus (1916a, trans. A. Hort) noted that "knot" or "bark blister" of olive might be attributed to a fungus. Orlob (1973) renders several pages of concise summaries, usually with the aid of Sprengel (1822), of passages relating to rot, canker, scab, fruit drop, mildew, rust, and other categories of disease; crops include olives, fig, grapevine, grain (wheat, barley), beans (not *Phaseolus,* a New World genus), fir, spruce, and ivy. There is attention to seed pathology, including which seeds keep well (millet and wheat) and which poorly (bean), and recognition that storage conditions are correlated with seed health. Although Theophrastus was scientifically naive from a modern perspective, he recorded doubts about the transmutation of cereals into weeds, noted some correlations between nutrient status and disease and between weather and disease, and even noted that in the case of rust, the emerging heads could contract disease from the leaves. He also noted differences among barley varieties in susceptibility to rust and repeatedly associated the differences with the readiness with which varieties shed moisture. For example: "Achilles barley . . . is apt to rust, whereas the eteocrithos barley is safe, since the ear bends . . . [and] water is shaken off" (Theophrastus 1990a, trans. Einarson and Link). Although these observations from Theophrastus may seem quite perspicacious, it was the opinion of Orlob (1973) that several diseases segregated in modern terminology (rust, smut, ergot, mildews) were viewed collectively in Hellenistic times. Ainsworth (1981) also comments on Theophrastus, with reference to Orlob (1973), and provides further translations and summaries from Theophrastus.

The most important of Greek writings addressing plants subsequent to Theophrastus is the *Geoponica,* probably compiled during the seventh century A.D. from Byzantine sources but containing works of some 30 authors living before A.D. 350 (Orlob 1973). This work addresses several topics pertinent to fungal plant pathogens. Good sanitary practices are recommended for storing grapes (removal of rotten fruit, clean facilities) and analogous practices for stored apples. The influence of humidity on *erysibei* (probably rusts in this context, but perhaps mildews) is recognized. The emphasis of the *Geoponica* is on disease control, including some commonsense advice on nutrition and the proper time for pruning, and the concept of seed treatment appears (although treatment by exposure to perforated sealskin may not have been very effective). Many remedies are based on superstition: burn three crabs (or fish) for protection against rust, transfer rust from crops by throwing laurel branches into the fields, the relationship of plant disease to the horoscope, and others.

The Romans wrote extensively on agriculture; most relevant to our topic are Cato (234–149 B.C.), Varro (116–27 B.C.), Virgil (70–19 B.C.), Columella (ca. A.D. 60), Ovid (43 B.C.–A.D. 17), and Pliny (A.D. 23–79), all reviewed and summarized by Orlob (1973), who relied on studies and translations by a number of authors. Cato's *De Re Rustica* mentions scab and fruit drop of fig, and interestingly, use of sulfur against caterpillars, but otherwise contains little on specific plant maladies. Varro's *Rerum Rusticarum Libri Tres* is most interesting for its invocation of miscellaneous deities to protect crops, including Robigo and Flora against rust, Ceres and Bacchus for protection of grain and grapes, respectively, Minerva for olives, and Venus for vegetables. Varro also made recommendations for seed treatment (olive dregs, or chalk, wormwood, etc.) and seed storage (cover legumes with a layer of ash and keep in jars). According to Orlob, boiled-down olive dregs (*amurca*) were a universal remedy against plant diseases and a forerunner of fungicides and insecticides.

Virgil's *Georgica* is probably the most famous of ancient writings on agriculture, largely because of its literary merit. It contains a few references to plant diseases, some of which are no doubt fungal in origin. Rust (*robigo,* recorded as a disease of the stem), seed treatment, and seed decay are specifically addressed. Columella's *De Re Rustica* is less literary but very comprehensive and more specific regarding plant diseases than the earlier Roman authors. He covers proper grain storage, seed treatment (remove defective seeds via flotation, or treat with leek juice), scab on grapes, fruit decay of pomegranate, and rust of grains.

Orlob (1973) relays a translation of part of Ovid's *Fasti,* impressive for its listing of plant maladies (translated as rust, smut, mildew) and the sacrifices and rituals necessary to avert them. Pliny was not an agriculturist, but his *Historia Naturalis* was encyclopedic and addressed several aspects of plant disease. Among his recommendations are use of *amurca*, dry and cool rooms for storage of fruit, and storage of apple fruits in wine. Modern plant pathologists know that stems of fruits can be infection courts for postharvest decay, so it is interesting that Pliny specifically recommended that when apples and pears were stored, "the stalks of all [be] smeared with pitch" (Pliny 1945, trans. H. Rackham). He also placed special emphasis on tree diseases ("worm infestation," "star blight," "indigestion," and "obesity" plus, more descriptively, carbuncle, scab, fungus on olive, and fruit drop). Pliny also mentions what could be interpreted as smut on oats, rust of grains, lodging of cereals, and seed treatment (be sure to carry a toad around the field the night before the field is hoed).

Finally, Orlob (1973) mentions Palladius, who was important by virtue of his influence on medieval writing. Palladius covered sanitation and pruning, transmutation (wheat to other grains), and rust (drive off causal mist by use of fire). Remedies deemed useful against various crop diseases included crucifixion of crayfish, burial of the head of an owl or the skull of a horse, use of hyena skin, etc.

Among the Greco-Roman remedies employed against fungal plant pathogens, it is interesting to note that seed treatment (with ashes, urine, wine, etc.) and fumigation (smoke of several sorts) were at least analogous to some modern approaches (Smith and Secoy 1975). One antirot treatment may have used arsenic. (Wijbenga and Hutzinger [1984], citing Pliny, treat the use of arsenic in vineyards as fact.) A wide variety of religious and magical practices (proper deployment of bare-breasted virgins, skulls of horses or asses, spells, burial of toads, etc.) were held to be effective against miscellaneous pests (Smith and Secoy 1975).

No account of plant pathology in ancient times would be complete without mention of the Robigalia festival. This recognition of the importance of plant disease and its incorporation into religious practice by the Romans was not entirely unique. Strabo (66 B.C.–AD 24), most famous as a geographer, noted a temple to Apollo Erythibius, named after *erysibe* (probably rust, according to Orlob [1973]), on Rhodes. Frazer (1955d) documented Strabo's reference as well as references of other ancient writers (including Homer, *Iliad* i 39) to "Mildew Apollo." But for the Robigalia, we have numerous, specific rituals documented in Columella and other works. As reviewed in Orlob (1973), Columella specified that Robigus (actually Robigo in Columella, who treats the deity as female) must be appeased by the sacrifice of a puppy. A skinned donkey head at the field margin was also recommended. The connection of rust with moisture was recognized by the necessity to appease Jove (Jupiter) lest excessive rains be sent. Ovid added (in his *Fasti*) sacrifice of a ewe, burning of incense, and libations of wine, and provided more detail on the origin of rust epidemics: a bad boy set a fox on fire and let it run through the fields. A common interpretation is that wheat rust was sent as a punishment for this cruel act (Kavaler 1965, Schumann 1991). The sacrifice of a dog appeased the Dog Star, which appeared at the time rust was most damaging. (Ideally the sacrificed dog should be of reddish color.) According to Pliny, the Robigalia was established by Numa (an early Roman king, ca. 700 B.C.). Buller (1915) gave a concise but well-documented account of the Robigalia, including the participants, the path of the ceremonial procession, and the culminating sacrifices. Buller remarked on a difference of opinion between Frazer and other authors on the significance of the red color of the dog. Frazer was said to be of the opinion that the red symbolized the color of the ripe grain, but others said of the rust. Rose (1922) was in the latter camp and was dubious about the connection with Sirius, the Dog Star.

According to Ainsworth (1976), "The Robigalia was the forerunner of the early Christian *litania major* (instituted by Pope Liberius [352–66]), which took place on the same date (St. Mark's Day) and involved a procession which followed the same route to the Milvian bridge but then returned to St. Peter's where mass was celebrated." This day, April 25, was also known as "Rogation Day" in the Christian calendar and was devoted to the blessing of crops (Schumann 1991). Frazer (1955d) supplies multiple references for ancient writings on Robigo and the Robigalia. Readers can also find concise and readable accounts of the Robigalia in Carefoot and Sprott (1967) and Large (1940).

Europe North of the Mediterranean

As would be expected, there is little information on how fungal plant pathogens were viewed in the context of cultural legacies of a largely preliterate Europe north of the Mediterranean. No body of knowledge exists analogous to that detailing the miscellaneous practices of the Roman Robigalia. What little can be adduced often comes from folklore generally acknowledged to have pagan roots. However, from this folklore we can surmise that fires (known as Lenten Fires, Easter Fires, Midsummer Fires, or Beltane Fires, depending on the season of their celebration) were sometimes used in rituals to preserve the health of crops (Frazer 1922, Tufnell 1924) and that dances or other ritual actions in the fields (preferably involving naked virgins) were performed to destroy agricultural pests (Murray 1955).

In an age lacking explicit knowledge of the microbial world, pestilence of crops was most often symbolized by a decidedly macroscopic pest, the mouse. Deities (and later, saints) with the mouse as an attendant symbol (e.g., the Germanic goddess Holda, and St. Gertrude of the Middle Ages) combined the functions of protecting crops against pests and protecting humans against contagion. These attributes were also characteristic of the Greek Apollo and so probably reflected an ancient and widespread phenomenon (Powell 1929). In general, neither the rituals (usually fire festivals) nor the deities or saints had plant protection functions as distinct from other functions. Health of livestock and of humans, fertility of crops, and protection of crops from bad weather, weeds, pests, and diseases were all facets of the same ritual or deity. Nonetheless, it is documented that the Lenten fire in Picardy was directed against smut (as well as against darnel and mice), and the charred wood from the Easter fire near Forchheim was buried on Walpurgis Day (May 1) to protect wheat from blight and mildew (Frazer 1922). These practices of near-modern European peasantry were undoubtedly more proximate to medieval ritual than to pagan practice, but the great antiquity of the fire festivals is presumed on a number of grounds, especially the indications that human sacrifice was integral to these rituals in pagan antiquity, e.g., the selection of sacrificial victims by various means (Frazer 1922). The corn spirits themselves might well require propitiatory sacrifice, as they were sometimes responsible for blackening the growing heads of grain as they glided through fields (Orlob 1971, summarizing Frazer).

Archaeological records add some depth to the sketchy information from folkloristic sources. Remains of ergot sclerotia (from both wild grasses and cereals), *Puccinia* spores, and spores of *Ustilago hordei* have been recovered in repeated instances from miscellaneous sites in northwest Europe, and both ergot and smut fungi were recovered from stomach contents of Iron Age "bog bodies" in Denmark (Dark and Gent 2001). The frequency of ergot increased considerably with the widespread cultivation of rye beginning in the Roman and medieval periods. Onnela et al. (1996) recorded fungal sclerotia associated with plant remains in a Viking age (A.D. 700–900) farming site in Finland. Dark and Gent (2001) extensively discuss the history of plant pathogens in northwestern Europe in relation to their "honeymoon" hypothesis (see Some Additional Hypotheses Regarding the Impact of Fungi in Ancient Times). Of particular significance was the related history of cultivation of hulled grains (emmer, spelt) and their putative resistance to pests and disease. These hulled grains were the characteristic cereals of the Neolithic, but their cultivation persisted in isolated pockets until much later, sometimes even into modern times. For example, the persistence of hulled grains in agriculture of the Iberian Peninsula was posited partially on the basis of putatively greater resistance to fungi compared to naked grains (Zapata et al. 2004).

Spread of Plant Pathogens via Transport of Seeds

Spread of seedborne plant pathogens was undoubtedly a fact of life in the ancient world, but like so many aspects of paleomicrobial ecology, much of the evidence is circumstantial. The primary forms of indirect evidence are the existence of very early navigation in the Mediterranean and adjacent areas, and the later strong development of transport of large amounts of agricultural produce by ship; the general pattern of the spread of agriculture; the resultant biogeographical pattern of domesticated crop species; and the archaeological evidence for the transport of arthropod pests in stored foodstuffs and seeds.

The general pattern of spread of agriculture from west Asia into the Balkans, then to central, western, and northern Europe was summarized in the Introduction. To this summary should be added here the studies of Colledge et al. (2004). In addition to revisiting the debates about the relative importance of movement of populations versus diffusion of techniques, Colledge et al. provide useful maps tracing the spread of various grains (hulled barley, emmer, einkorn) throughout the eastern Aegean, including islands, from ca. 10,000 B.C. to 6500 B.C. Hulme (2004) states that many species of cereals and legumes were introduced to Mediterranean islands in the Neolithic.

The literature on early navigation is abundant, and it is not necessary to review it in detail here. However, it is useful to sketch the outlines of research results. There is ample circumstantial evidence that sporadic seafaring was already occurring in the Pleistocene, although no physical evidence of Pleistocene seafaring or credible depictions of Pleistocene seacraft exist (Bar-Yosef 2002, Bednarik 2003). The earliest archaeological remains of seacraft reflect maritime trade between the southern Mesopotamian Ubaid culture and the Persian Gulf in the fifth and sixth millennia B.C. (Carter 2006). However, findings dated starting from around 11,000 B.P. of obsidian from islands of the Aegean (mainly Melos), plus findings of certain stone tools dating from even earlier, argue strongly that specific types of stone valuable for toolmaking were transported long distances, and often from islands to the mainland or other islands (Bar-Yosef 2002, Bednarik 2003, Perles 2001). Bednarik (2003) concentrates on very early navigation in the Southeast Asian archipelago but argues for extremely early dates in some parts of the Mediterranean as well. Such early dates greatly precede the Neolithic, and it is highly plausible that primitive but regular sea transport was well established in coastal regions before the advent of agriculture. Bellard (1995) proposes on the basis of radiocarbon-dated human remains that some of the Balearic Islands were colonized as early as around 6000 B.C. When agriculture did arrive in a given region, it often spread along coasts and rivers (Bar-Yosef 2002).

By the Neolithic, sea transport was well developed in the Cyclades. Sea routes established between the Cyclades and early Minoan Crete persisted into the early Bronze Age (Agouridis 1997, Betancourt 2003, Broodbank 1993, Hayden 2003, Petrakis 2004, Şahoğlu 2005). The earliest representations of ships (petroglyphs, pottery shards) depict longboats powered by oars and imply a lack of sails (Broodbank 1993). Schwartz (2002) writes that the earliest fragments of reed boats are from a sixth millennium B.C. site in Kuwait. Analogous fragments from the Euphrates in southeast Anatolia date from 3800 B.C., and pictographs on clay tablets from fourth millennium B.C. Mesopotamia show reed boats. Riverine transport by oared longboats also developed early in predynastic Egypt (Berger 1992). In all likelihood, seed grain was among the commodities transported by such early ships in the later eighth or early seventh millennium B.C. (Broodbank and Strasser 1991, Davis 1992). Thus we can surmise that transport of seedborne pathogens also began extremely early.

Sea transport of plant materials in the Bronze Age is well documented via the archaeology of ancient shipwrecks. The Ulu Burun shipwreck (Turkey, late fourteenth century B.C.) contained seeds of many plants, including wheat, barley, fig, grape, and *Carthamus* (Haldane 1991, 1993). Twigs and brush were transported as dunnage packed between cargo and hull (Haldane 1991). Recovery of cereal grains and pulses probably reflects shipboard diet (Haldane 1993). The principal cargo of the Ulu Burun ship seems to have been terebinth resin (Haldane 1990), valued throughout the Bronze Age and later for its aromatic properties and used as incense, in cooking, and as a flavoring and preservative for wine (McGovern 2003).

Not only were ships carrying seeds and other plant products from very early times (Neolithic, early Bronze Age), but by the early Iron Age, even before the invention of coinage, market-type trade in grain, olive oil, and wine was well developed in western Asia and the Mediterranean, a conclusion based largely on the analysis of ancient shipwrecks (Temin 2003). Such trade increased in Greco-Roman antiquity. Pollen and phytoliths are increasingly being used in archaeological analysis of shipwrecks and provide evidence of Persian and Roman cargoes of cultivated grasses and plants used as spices (Gorham and Bryant 2001). Evidence of plant or plant-derived cargoes transported in amphorae (olive oil, wine) is common for the Roman period (e.g., McCann 2001), and transport by sea of large quantities of grain is well documented (e.g., Rickman 1971, 1980). Evidence for plants used as food for the crews (grain or lentils, etc.) is less direct, but transport of such items is inferred from ship storage facilities and containers (McCann 2001).

Some of the strongest circumstantial evidence for seagoing transport of seedborne phytopathogenic fungi comes from the recovery of arthropod pests. It can reasonably be assumed that seedborne fungi can survive long periods of transport and storage at least as well as arthropod pests: "Seedborne pathogens, particularly those which are not wholly superficial, persist at least as long as the useful life of the seed" (Maude 1996). Haldane (1991) notes the recovery of weevils and bruchid beetles from the Ulu Burun shipwreck remains. Panagiotakopulu and Buckland (1991) present further evidence for human transport of weevils and bruchids in stored products in the Bronze Age Aegean. Buckland (1981) reviews dispersal of insects, principally Coleoptera, in stored products from archaeological sites in Egypt and Roman Britain. Pests of stored cereal grains and legumes can be demonstrated to have been distributed well distant from their natural ranges during antiquity.

It should be noted that fungal hyphae have indeed been documented in association with the plant remains from ancient shipwrecks, but the concern has been that they are invasive (Gorham and Bryant 2001). Unless spore type or other evidence demonstrates that microbes associated with archaeological remains are contemporary with the excavated material, they must be considered as possibly invasive at a later date. Despite these reservations, the strong circumstantial evidence for early movement of seeds, the evidence from archaeoentomology, the occasional archaeological find of fungi that can act as seedborne pathogens (e.g., several genera in Stewart and Robertson 1968), and the fact that ancient peoples attempted to treat seed with ashes, olive dregs, or other substances indicate that seedborne plant-pathogenic fungi were moved about the ancient world and had an impact on early agriculture.

Indeed, some modern plant pathologists have factored into their hypotheses the possibility of early seedborne transport. Seedborne *Ascochyta* may have been moved from the Balkans to the Levant (Abbo et al. 2003). Infected seed was a possible mechanism for the movement of *Mycosphaerella graminicola* from its center of origin into Europe (Banke et al. 2004). And Dark and Gent (2001) propose that the movement of seedborne fungal diseases in the late Iron Age and Roman periods was a factor in the increased incidence of plant disease in Europe.

Fungi as Agents of Rot on Wood and Fabric

The deterioration of structures and fabric containing cellulose or other fibers was readily noticed and recorded by ancient peoples. As with plant pathogens, instances of rot recorded in the Bible have been scrutinized. The putative microbial agents have varied with the scientific climate of the scrutinizers, and it is probably not surprising that *Stachybotrys* species (the current archvillains of indoor air fungi and toxic building syndrome) have now entered the lists! "Tutankhamen's curse" has been attributed to fungi growing on ancient grave goods. Understandably (and more realistically), most writing about ancient fabric and wood focuses on the preservation of the artifacts from attack by contemporary microbes, and the presence of any ancient microbial remains has been of far less interest.

Documentation: Fungi as Agents of Rot on Wood and Fabric

Tsara'at = Mold

The ancients certainly knew that wooden structures and fabric would rot, and modern writers have attempted to identify the causes of such rots. Rots of garments are mentioned in Leviticus 13:47–59, which refers to "leprosy" of various garments. Contenders proposed by various scientists include *Microsporum gypseum* (also an agent of ringworm in humans and animals), *Chaetomium globosum,* various species of *Penicillium* and *Aspergillus, Memnoniella echinata,* and *Cladosporium herbarum* (Moldenke and Moldenke 1952). Heller et al. (2003) maintain that the word in Leviticus 13 above and 14:33–57 usually translated into English as "leprosy" bears no relation to that medical condition. Although it may be manifested as a scaly skin condition, the word, *tsara'at,* is also applied to conditions of garments as above and the walls of houses. The authors suggest that one of the manifestations may have been caused by a *Stachybotrys* sp. and that *tsara'at* is better translated as "mold." Koch (1983), also in reference to Leviticus, mentions *Aspergillus, Penicillium, Chaetomium,* and other fungi as candidates and points to the possibility of contamination by mycotoxins.

A series of reports, summarized in somewhat melodramatic fashion by Janińska (2002), have connected the deaths of several eminent persons excavating ancient tombs with exposure to mycotoxin-producing fungi (the "Tutankhamen's curse" phenomenon). A more concise and constrained account is found in Yiannikouris and Jouany (2002). Sledzik (2001) gives a more detailed account and cites (but does not necessarily endorse) an opinion that "fungal spores may be a source of infection even when thousands of years old."

Conservation of Artifacts

Probably the majority of publications on the presence of fungi in archaeological remains pertain to conservation of the remains themselves rather than the study of fungi belonging to the time period when the artifacts were made. Clearly, many records of fungal rot in wood, fabric, and associated materials are the results of fungi that are intrusive. Blanchette (2000) reviews several instances of degradation of ancient artifacts, primarily wooden, by microbes. It is difficult to date the timing of such decay, which "could have been introduced at any time over the past centuries with degradation occurring whenever conditions were favorable" (Blanchette 2000). The mummy of Ramses (Rameses) II, restored in Paris in the mid-1970s, was much contaminated by many species of modern fungi (Rollo and Marota 1999). The *Journal of the American Institute for Conservation* is a good source for information on multiple aspects of the degradation of collections, art, and artifacts by fungi. See also the general reviews of Ciferri et al. (2000) and Florian (1997). Hawks (2001) gives a concise history of the chemicals used for insect and mold control on such artifacts.

Interestingly, fungal deterioration of artifacts also plays a role in determining their authenticity (Museo d'Arte e Scienza n.d.):

> An absolutely certain indication of authenticity is provided by carbonized mould fungi, which develop on edible burial gifts. These fungi expand radially in an irregular way. In the course of centuries a microorganism (Micrococcus carbo) converts the fungus material into carbon. Under a magnifying glass they appear as a crystalline mass, whilst the black stains forgers like to make by spraying black paint on the surface show up as round, smooth-surfaced spots.

Human and Animal Pathogens

9

Fungal infections are uncomfortable and often deforming, and it is not surprising that ancient writings refer to fungal disorders of the skin (ringworm), thrush (a yeast infection), and mycetoma (causing swelling and deformities, usually of the foot). Prior to written accounts, some such disorders, such as skin diseases, were recognized in myths. Putatively ancient folk cures for ringworm lingered into modern times, when they were documented by folklorists. It is highly probable that some ancient skin diseases translated from myths or the Bible as "leprosy" were not that condition (Hansen's disease, caused by *Mycobacterium leprae*) but rather were caused by fungal agents of ringworm such as *Microsporum* or *Trichophyton*. Fungal skin diseases often subjected their hosts to radical social stigma, and especially so when confused with true leprosy.

Documentation: Human and Animal Pathogens

Ringworm and Other Nasties

One of the most frequently documented fungal diseases in ancient times is ringworm. "The first recorded reference to a dermatophyte infection is attributed to Aulus Cornelius Celsus, the Roman encyclopedist, who in his 'De Re Medicina' written around 30 A.D., described a supperative infection of the scalp that came to be known as the kerion of Celsus" (Ajello 1974, citing Rosenthal 1961). The term Celsus actually applied, *porrigo*, is now obsolete but was repeated by Pliny and by dermatologists up to the nineteenth century (Ainsworth 1993a).

There are numerous references to ringworm in the writings of historians, classicists, and folklorists. Fungal infections were among the skin diseases treated in Roman practice with magnesium salts (Cruse 2004), and they may have spread when people used the public Roman baths

(Nutton 2004, citing Pliny). Leviticus 13:29–37 contains a plausible reference to ringworm (favus) in describing sores on the head or beard (Mark 2002). Based on context in Homer and similarities to modern Greek usage, the word μολοβρός as it appears in the *Odyssey* is taken as baldness linked to ringworm disease (Coughanowr 1979).

Ringworm was also well known to the Anglo-Saxons (Saunders 1965, citing Bonser 1963). Mooney (1887), a nineteenth-century collector of purportedly ancient Irish folkways, noted that in Ireland and Scotland, a common cure was to rub the afflicted area with gold or silver coins. Another of several cures was to rub the area with the blood of a black cat, "and in some houses there are cats whose ears have been cut away piecemeal for this purpose." The Irish name for ringworm was *teine-d·iad·a* or "divine fire," and Mooney noted that red and black were the principal colors of Irish mythology. Mooney put his collected folkways in the context of the "Fairy Influence," implying pagan origins. In Europe as well as the Mediterranean, ringworm was sometimes confused with leprosy, as documented below. (See also The Anglo-Saxon Charms in Chapter 13.)

Dalby (2004), citing several classical texts, recounts the Greek myth of the daughters of Proitos, noting that "leprosy covered all their flesh, and their hair dropped from their heads, and their smooth scalps were bared." Although hair loss, especially of eyebrows, eyelashes, and body hair, can occur with leprosy (i.e., Hansen's disease) (Mark 2002), it is equally or more likely that severe hair loss from the scalp is from infection with dermatophytes causing ringworm of the scalp (tinea capitis). *Microsporum* species (such as *M. gypseum*, Figure 8), well documented in Mediterranean countries, are a cause of tinea capitis (Kwon-Chung and

FIGURE 8
Microsporum gypseum, an agent of ringworm. This globally distributed species was likely among the agents of ringworm that produced symptoms the ancients confused with leprosy. (From Howard et al. 2003; reprinted by permission of the authors and Taylor & Francis.)

Bennett 1992). According to Kane (1997), "from Biblical times through the Middle Ages, many people were incorrectly sequestered as lepers due to the confusion between favus and leprosy" (favus is the condition in which hair becomes encrusted with follicular pus in dermatophyte infection). Lewis (1987) held an even stronger opinion on "leprosy" in Biblical times: "The Biblical leprosy was not what we call leprosy;" it covered a number of conditions, including fungal infections, but never the condition caused by *Mycobacterium leprae*. Monot et al. (2005) present evidence favoring an ultimate east African origin of *M. leprae* but withhold judgment on the timing of its appearance in Greece and the Aegean. However, they note opinions that the troops of Alexander the Great carried the disease from India. Mark (2002) argues that the disease was transported to the Mediterranean about 400 B.C. in cargo ships, not by Alexander's army. The *Oxford English Dictionary* (1971 ed.) records the Hebrew and Greek words translated as leprosy but states that these words actually covered a variety of skin diseases. Graves (1960) notes other Greek myths that mentioned leprosy. In "The Conquest of Elis," Graves refers to "Lepreus, the son of Caucon and Astydameia, who founded the city of Lepreus in Arcadia (the district derived its name from the leprosy which had attacked the earliest settlers)" and states that the form of leprosy denoted was *vitiligo* (which Graves, in contrast to modern medicine, attributed to poor food). In "Auge," Graves tells the story of Teuthras, who killed a boar in the temple of Artemis in spite of the boar's pleas for its life. As punishment, Artemis inflicted leprous scabs on Teuthras. In his chapter on Elis, Graves also states that true leprosy (Hansen's disease) did not arrive in Europe until the first century B.C. Graves (1966) presents, in conjunction with the healing powers of specified deities, a synopsis of the various types of "leprosy" in archaic Europe. It seems probable that at least some forms of disease rendered as "leprosy" in translation were not incited by *M. leprae* but were of fungal origin, i.e., ringworm.

Reference to "leprosy" in animals in Greco-Roman veterinary practice (Moule 1990) might also be attributed to ringworm. Naturally, there is little direct reference to fungi in ancient veterinary medicine (but see preceding discussion of heaves in horses). However, the Romans especially were avid beekeepers and acute observers of diseases in the hive. Columella (1941b, trans. Forster and Heffner) in *De Re Rustica II* mentioned several diseases, identified by the translators as foul brood (caused by bacteria) and dysentery (caused by a microsporidian), but no disease specifically identified as fungal. Virgil (2002, trans. K. Chew) wrote both prolifically and poetically about bees and about their diseases. Based on symptoms, Chew suggests, in addition to the bacterial and microsporidian agents, the fungi *Ascosphaera apis* (agent of chalk brood) and *Aspergillus flavus* (agent of stone brood).

There were other troublesome fungal infections. "Possibly the earliest record of a mycotic infection is that in the Indian Atharva

Veda [ca. 2000–1000 B.C.] of mycetoma of the foot" (Ainsworth 1986). Ainsworth (1993a) gives the Vedic name for the condition as *padaavalmika* ("foot anthill"); a similar commentary can be found in Kwon-Chung and Bennett (1992). The *samanu* disease of Babylonian texts was probably also mycetoma (it was treated with, among other things, a plaster of lees of beer) (Wilson 1994). Thrush, a yeast (mostly *Candida albicans*) infection common in infants, was documented in Hippocratic writings (fifth century B.C.) as well as in the commentaries of Celsus (Ainsworth 1986, 1993a).

Direct documentation of fungal infections in an archaeological context is rare. Interestingly, given the early accounts of what was purportedly mycetoma (see above), Lowenstein (2004, citing Hershkovitz et al. 1992) lists Madura foot, a type of mycetoma, from Israel (ca. A.D. 300–600). The condition is rare in present-day Israel, and the case is suspected to have migrated to the location of recovery via trade routes of the time.

Fungi and Healing Rituals

The ancients drew no clear line between magic and medicine, and it is therefore relevant to provide examples of magic involving fungi or products of fungi. One ill-portending fungus (apparently seen under a stone and perceived as harmful to persons) was an object of fear against which spells and sacrifices were invoked, as recorded in Akkadian tablets (Caplice 1974, Reiner 1995). Libations of beer were part of the spell to avert the evil effects of the fungus, and beer was used for other similar purposes, as well as for consuming herb potions and anointing the body with magic powders (Reiner 1995). Offerings of wine and bread to the gods were standard when the Druids harvested sacred herbs (Freeman 2006). The stinkhorn, *Phallus impudicus,* is mentioned by Reiner (1995) in a context pertaining to the semantics of Sumerian and Akkadian names for plants, but although it is native to the region, its significance to Mesopotamians is unclear. Because of its obvious phallic connotations, the stinkhorn long featured in homeopathic practice and the allied medieval doctrine of signatures (Molitoris 2002). See also Fungi in Ancient European Folklore for uses of wine, mead, and the like in spells and magic.

Environmental and Ecological Roles of Fungi

Certain fungi are consistently associated with specific substrata and are therefore reliable indicators of a given environment. Most notable in this regard are some genera of dung fungi, whose presence and/or abundance is correlated with herbivores and abundant fodder, usually grasses. Other substrate-specific fungi degrade hair, hooves, and horn. Much of the literature using fungi as indicative of a given substrate pertains to times before the Neolithic (i.e., Paleolithic megafaunal extinctions), but some researchers have used dung fungi to chart human hunting and herding activities during the Neolithic.

Other fungi of potentially great ecological importance are the pathogens putatively responsible for the decline of elms during the Neolithic in western Europe. As yet there is no consensus on whether or not a species of *Ophiostoma* (formerly called *Ceratocystis* in the case of elm disease) was primarily responsible for this decline, but given the impact of Dutch elm disease on modern urban trees in Europe and America, there is much interest in the analogous events of the Neolithic.

Documentation: Environmental and Ecological Roles of Fungi

Coprophilous Fungi and Archaeobotany

Fungi recovered in archaeological contexts have been used as indicators of the prior environmental status of the archaeological site. Of particular interest are ascospores of *Sporormiella* species, which occur principally on herbivore dung. Spore-forming *Gelasinospora, Coniochaeta,* and *Chaetomium* also are frequently coprophilous. In conjunction with pollen analysis and study of charcoal remains, spores of these fungi have been used to estimate herbivore abundance and grazing and to infer use of fire by humans to promote ungulate density in British woodlands ca. 4300 B.C. (Innes and Blackford 2003). *Sporormiella* spores have been repeatedly used worldwide in efforts to trace megafaunal extinctions (Comandin and Rinaldi 2004). Spores of *Sporormiella, Podospora, Chaetomium, Cercophora, Glomus,* and other genera were used to supplement other evidence in deducing a grazed grassland environment surrounding a settlement of Roman times, and spores of several of these genera were used to characterize the environment during the Bronze Age at another site, both in the Netherlands (Buurman et al. 1995, van Geel et al. 2003). Ascopores of Sordariaceae served as indicators of herbivore activity at a Neolithic site in northeastern Italy (Pini 2004). Burres (1995) describes hair- and keratin-degrading fungi from late Pleistocene scat.

What Killed the Elms?

Depositions of pollen, plant remains, and remains of an insect vector have at times been used to infer the presence of fungi. Perry and Moore (1987) conclude that Neolithic elm decline in northwest Europe, previously attributed to various causes, can best be explained on the basis of disease caused by *Ceratocystis* sp.; this conclusion is contested, at least for Britain, by Robinson (2000), who reviews the literature on elm decline and Neolithic woodlands. Clark and Edwards (2004) are more supportive of the hypothesis that the decline was induced by fungi. Conclusions are tentative because fossil remains of the scolytid vector in the pertinent context are scarce. Moe and Rackham (1992), as an aside in their discussion of the possible association of *Ceratocystis* with Neolithic elm decline, note an isolated population of diseased elms on Crete and posit that both the *Ceratocystis* and the elms on Crete were remnants of an earlier postglacial distribution. (*Ophiostoma* is the generic name applied to fungal agents of this elm disease in more recent publications; see Clark and Edwards [2004] for a review of recent literature.)

Ancient Fungi Preserved in Glacial Ice or Permafrost

Few recent scientific developments are as exciting to mycologists and microbiologists as the possibility that ancient microbes have been naturally cryopreserved in glacial ice or permafrost and can be revived and cultured. Even if such microorganisms are no longer viable, their DNA might be sequenced, with profound repercussions for our understanding of the rate of microbial evolution. Given that many microfungi are cosmopolitan and can be transported long distances by winds, microfungi recovered from ice cores in Greenland or Siberian permafrost may well represent fungi common in ancient temperate environments. Some researchers have claimed that fungi (or bacteria) several thousand years old have been recovered from natural cryopreservation into active growth on artificial media. Other scientists have isolated and identified metabolites produced by such fungi. There have even been analogous claims for microorganisms hundreds of thousands or even millions of years old. Some of the more extreme claims have aroused suspicion that the "ancient" microbes may actually be modern or near-modern contaminants; as a consequence, much is now written on safeguards to protocols pertinent to these studies. Of particular interest are studies of fungi associated with the Tyrolean Iceman, whose case provides an excellent example of the possibilities, pitfalls, and controversies associated with naturally cryopreserved fungi.

Documentation: Ancient Fungi Preserved in Glacial Ice or Permafrost

Ice and Permafrost

There are several claims that fungi of great antiquity have been detected in or recovered from ancient ice or permafrost. In most of these instances, the polymerase chain reaction (PCR) was used to amplify and sequence portions of the genome typical of specific groups of fungi. For example, ice cores 2,000–4,000 years old from north Greenland yielded sequences diagnostic for rusts, basidiomycetous yeasts (*Leucosporidium* and *Rhodotorula*), ascomycetous yeasts, pyrenomycetes, and loculoascomycetes (Willerslev et al. 1999). This same group of researchers investigated fungi in Siberian permafrost from the Holocene and late Pleistocene (and even early Pleistocene) and detected sequences typical of modern fungal genera, including *Aureobasidium, Blumeria, Cryptococcus, Hormonema, Melampsora, Mortierella,* and *Sordaria* (Lydolph et al. 2005). Analogous techniques detected fungi in Greenland ice cores ranging in age from 300 to 140,000 years B.P. (Ma et al. 2000). Some isolates were viable and morphologically identifiable, including common small-spored fungi such as *Penicillium, Cladosporium,* and *Ulocladium*. One fungus, *Cladosporium cladosporioides,* was isolated from cores dated to 700 and 4500 B.P., presenting the possibility of studying the microevolutionary history of the species.

In analogous studies, Vinokurova et al. (2005) were able to obtain the metabolite mycophenolic acid from *Penicillium* strains isolated from ancient (mostly 30,000–4000 B.C.) permafrosts, paleosols, or glacial ices, and Kozlovskii et al. (2003) document production of agroclavine and epoxyagroclavine by *Penicillium citrinum* isolated from 1.8 million-year-old(!) permafrost from the Russian arctic. Kochkina et al. (2001) claim recovery of viable fungi of a wide range of taxa from permafrost 5,000 to 10,000 years old and (more controversially) recovery of microorganisms from permafrost dated two to three million years B.P. Prominent were *Alternaria, Aspergillus, Botrytis, Chaetomium, Cladosporium, Geotrichum, Penicillium,* and other common genera. Although most studies involve polar ice, it appears that such ice entraps fungi arriving from a great distance. Rogers and Castello (2001) cite multiple publications for that conclusion and also give a concise literature review on characterization of fungal and bacterial communities from ice cores.

The Iceman

Much interest has also focused on microbial remains associated with the Tyrolean Iceman (Ötzi), radiocarbon-dated to ca. 3350–3100 B.C. Several publications address fungi taken from various parts of Ötzi's body or attire. Rollo and coworkers (1994, 1995, 1999) describe studies in which fungal DNA was isolated, including the basidiomycetous yeast *Leucosporidium scottii* and two ascomycetes. Rollo et al. (2002) later extracted DNA sequences diagnostic for two phyla (Ascomycota and Basidiomycota) and classes (Urediniomycetes, Heterobasidiomycetes) from the corpse's digestive system. Haselwandter and Ebner (1994) refer to viable isolates of *Chaetomium* and *Absidia* recovered from the Iceman, but this result has been disputed (Gams and Stalpers 1994). Marota and Rollo (2002) give a capsule history of mycological studies on the remains of the Iceman, placed in context with other studies in molecular paleontology.

Caveats to the Study of Ancient Fungal DNA

Pääbo et al. (2004) pay scant attention to fungal DNA on ancient materials, seeing it as essentially only a contaminant. Similarly, Rollo and Marota (1999) cite studies on an Egyptian mummy in which many modern microbial contaminants were isolated. Gorham and Bryant (2001) mainly concern themselves with fungal hyphae as modern, intrusive elements in botanical remains from ancient shipwrecks, and Blanchette (2000) expresses similar concerns. Clearly, rigorous safeguards, not to mention some skepticism, are warranted when amplifying "ancient" fungal DNA. Cooper and Poinar (2000) and Willerslev and Cooper (2005) discuss some of the challenges presented by possible contamination of ancient substrata with modern microbial DNA and suggest appropriate safeguards against erroneous conclusions.

Ancient Images of Fungi

Mycologists, historians, archaeologists, folklorists, and others have taken a keen interest in ancient depictions of fungi. These images occur as petroglyphs, on ancient ceramics, as stone carvings, and as engraving or embossing on metal. Since humans produced images before written language, the oldest putative images of fungi may be quite ancient. Some of the images seem conspicuously fungal, usually mushroom-shaped, whereas others are only plausibly representative of fungi. Interpretation of most such images has varied widely, since virtually none are accompanied by written documentation by the original artist. A very few, such as the depiction of *Lactarius deliciosus* on a wall at Pompeii, are relatively unambiguous as to the taxon represented. The vast majority, however, have been interpreted according to the (often conflicting) biases of modern historians, anthropologists, and others, and some skepticism is in order when reviewing such interpretations. (Does the "hugely dilated" eye on the centaur truly indicate that the mushrooms at his feet are hallucinogenic? Are the shapes on the Palette of Narmer really giant *Psilocybe* mushrooms, or are they just Egyptian tote bags?) In several instances, interpretations of images are based on detailed, often highly scholarly analysis of components of myth, ritual, and philology. Several of the investigators, including some of the most scholarly, have had a strong personal interest in experience with psychoactive substances.

Documentation: Ancient Images of Fungi

Phalli and the Dancing Myco-Shamans

Compared to images of deities, monarchs, ships, weapons, animals, or even higher plants, ancient images of fungi are rare, but they do exist. The oldest may be those from ca. 7000–5000 B.C. from Tassili in the Sahara

Desert, Algeria (Samorini 1992), which have been interpreted as shamans wearing mushroom costumes and were held to represent hallucinogenic fungi. The figures as reproduced in Samorini (1992), who visited similar sites in Libya, Chad, and Egypt, render this interpretation plausible but not certain. Samorini also cites publications on early mushroom representations (possibly *Amanita muscaria* from proportion and ornamentation) in petroglyphs from Siberia ("local neolithic") and mushroom motifs in prehistoric petroglyphs in the Kamchatka peninsula. These figures were interpreted as representing shamanistic activities. Samorini concisely references Kaplan's (1975) theory that mushrooms, possibly *A. muscaria*, are depicted in Bronze Age cave art from Sweden. Several of the figures reproduced by Kaplan involve mushroom-shaped structures, placed in the middle of what appear to be double-prow longboats, that look considerably more like sails than mushrooms; other figures more plausibly represent something fungal, although Kaplan notes that cult-hatchets have also been suggested. Kaplan, like Morgan (1995), remarks on the Swedish custom of tossing mushrooms into bonfires (cf. Frazer 1955e).

Rivaling in antiquity the figures from Tassili are those discussed by Gimbutas (1982, p. 220):

> *Dwelling areas at Vinča [Balkans, just east of Belgrade] yielded a number of mushrooms carved out of light green rock crystal . . . [Figure 9]. Mushrooms are universally known as aphrodisiacs, and the swelling and growing of a mushroom must have been noticed by the Old Europeans causing it to be compared to the phallus. The fact that mushrooms were carved out of the*

FIGURE 9
Stone carving of a mushroom from the Chalcolithic, Vinča, Balkans. (From Gimbutas 1982; reprinted by permission of University of California Press.)

best available stone speaks for the prominent role of the mushroom in magic and cult. The Indo-Europeans in the days of the Rigveda made their miraculous Soma drink from the fly-agaric (Wasson 1971), and it is possible that the Vinča mushrooms were connected with intoxicating drinks; at all events they are imitations of phalli . . . The shape of the mushroom or phallus occurs frequently in sculptural art as a human cap on figurines, and a phallic form can be inferred in the beautiful Butmir vases which are decorated in running spirals . . .

Earlier in this same chapter, Gimbutas had written, "The whole group of interconnected symbols—phallus (or cylinder, mushroom and conical cap), ithyphallic animal-masked man, goat-man and the bull-man— represents a male stimulating principle in nature without whose influence nothing would grow or thrive. This family of symbols goes back in its origins to the early agricultural era." She states that the Vinča site represented the Chalcolithic, ca. 5500–3500 B.C. Although one of the figures referenced in Gimbutas does have phallic connotations, other figures strongly resemble agarics more than any other common organism or object. The antiquity of the objects is interesting in light of claims in Gimbutas (1999) for the great antiquity of Baltic folk beliefs in general, and especially those regarding fungi, although the mushroom stones from Vinča are Balkan, not Baltic, in provenance. However, "there is evidence in Latvia to prove the existence of idols made of two slabs of stone in the form of a mushroom" (Urtāns 1992). (See Fungi in Ancient European Folklore.)

It is possible that a simplified graphic for mushroom was a component of ancient writing. The markings sometimes referred to as "Old European script" have been proclaimed as constituting a system of writing preceding even that of the Sumerians and possibly connected to much later linear scripts such as Linear A and the Cypriotic script of the first millennium B.C. (e.g., Gimbutas 1999; Marler 2001). Just as the V mark was putatively used to indicate the female pubic region and fertility, the inverted V (Λ) of Old European "was used to indicate the cap of sacred mushrooms" (Russell 1998, citing Gimbutas 1991). Russell is perhaps even more idiosyncratic than Gimbutas, but this latter notion does bear an interesting resemblance to the ideas of Toporov (1985), who documents the semiotic correspondence between invaginated and protuberant agaric caps and the vagina and phallus, respectively, in old Slavonic language and folklore. Although the system of signs termed Old European has been intensively studied for symbolic content (e.g., Merlini 2006), the reader should be aware that the conventional academic perspective has been quite dubious about its constituting a genuine system of writing (e.g., Meskell 2000).

Also noteworthy are the illustrations reproduced by Morgan (1995) from Egyptian and Hittite sources. The first, from the Palette of Narmer (an early Egyptian ruler), shows very large, mushroom-shaped objects with curiously limp stipes (stalks) slung over the shoulder of a person who "appears to carry psilocybin or related fungi" (Morgan 1995). The second is a series of three figures from Alaça Huyuk, Turkey (second millennium B.C.), each of which consists of a pair of mushroom-shaped figures with one or more human features (eyes, arms, etc.).

Greco-Roman Images

Most ancient imagery of fungi from the Mediterranean region is from the Greco-Roman period. Samorini and Camilla (1994) summarize the representations from Greek art. Several of the images have been reproduced, including one (Figure 10) also used as a cover illustration for one of the two volumes of the Penguin reprint of Graves (1960) and also reproduced by Morgan (1995) and Ruck et al. (2001). A fresco from Pompeii very probably portrayed the edible *Lactarius deliciosus* (Buller 1915).

Graves (1960) comments at length on an Etruscan representation (on a mirror) of a fungus pertinent to the myths of Ixion and Sisyphus. Reproductions of the mirror can be found in Morgan (1995), Ruck et al. (2001), and Wasson and Wasson (1957). Graves (1958, "What Food the Centaurs Ate") discusses the Etruscan image and its significance in even greater detail, noting that a mushroom appears between the hooves of a dying centaur on a Greek vase ("clearly a *panaeolus*"—a genus with some hallucinogenic properties). Several other mushrooms are pictured on the vase, and the pupil of the centaur's eye is "hugely dilated." Both the generic identification of the mushroom and the comment about the

FIGURE 10

Greek goddesses with mushrooms. This sculpture has been often reproduced in support of the theory that Greek religion was influenced by the use of psychoactive fungi. (From the Louvre, © Réunion des musées nationaux/Art Resource.)

degree of dilation of the centaur's eye are, objectively, more reflective of Graves's perspective than verifiable data, but Graves's conclusions are at least plausible. Morgan (1995) reproduces an image of an Attic vase that portrays Perseus with mushrooms and presents an eclectic array of other ancient images demonstrably or putatively depicting fungi from Persian and other ancient sources. Ruck et al. (2001) reproduce numerous images, including a convincing trio of mushrooms on a platter, held by a man said to be a priest attending the sacrifice of a pig by Heracles (Figures 11 and 12). The authors speculate that the mushrooms were *Amanita muscaria*,

FIGURE 11
The Lovatelli Urn. The scene reputedly shows sacrifices performed by Heracles. (From Lovatelli 1879)

FIGURE 12
Detail from the Lovatelli Urn: Platter of mushrooms. (From Lovatelli 1879)

with "three being the approximate dosage." Some other Greco-Roman images are mentioned in passing in Fungi, Philology, and Mythology.

Although not produced in Greco-Roman times (the fresco is from thirteenth-century France), the illustration on the back cover of Allegro (1970) is highly relevant because Allegro strongly contends that it reflects a tradition from classical antiquity. It seems to depict the tree of life in the Garden of Eden as several *Amanita* fruiting bodies arising from a common locus. Each of the "mushrooms" clearly possesses a mushroom-shaped cap, a distinct stalk, and dots on the cap. A serpent twines about the central mushroom, and Adam and Eve flank the mushroom cluster. Photographs of the same fresco can also be found in Morgan (1995) and Ramsbottom (1953). (See the critique of Allegro [1970] in Fungi as "Entheogens.") To his credit, Allegro supplies a footnote regarding alternative interpretations of the figure.

Agarics and Huns

Among the seemingly most explicit but nevertheless mysterious ancient images are the mushroom figures embellishing the lugs of bronze cauldrons manufactured by the Huns (Figure 13). Very little is known about the language, religion, or ethnicity of the Huns, in part because Attila the Hun was a great assimilator of other peoples and customs (especially Germanic and/or Iranian) in furtherance of his designs of conquest. There are some indications that Turkish words were especially impor-

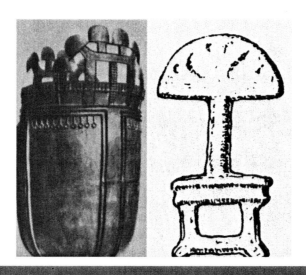

FIGURE 13

Left: A cauldron made by the Huns and decorated with characteristic mushroom-shaped lugs. Right: A fragment of a mushroom-shaped lug. (Both from Maenchen-Helfen 1973; reprinted by permission of University of California Press.)

tant, and there is evidence of shamanistic elements in Hun practices (Maenchen-Helfen 1973). The mushrooms on the cauldron lugs in the figures reproduced by Maenchen-Helfen are not only explicitly agaric in outline, but also the pileus (cap) is clearly separate from the stipe (stem), and on at least one cauldron the stipes are clearly shown as centrally attached to the underside of the pileus. The pileal surface is usually plain but occasionally ornamented with a single disk, or with lines, or occasionally with a variety of other shapes. Some individual mushrooms sit on pedestals (e.g., Figure 49 in Maenchen-Helfen, reproduced here as Figure 13 [right]), and a very similar "icon" of a mushroom on a pedestal appears in a rock picture that also includes a number of cauldrons (Figure 53 in Maenchen-Helfen). The cauldrons were used in ordinary cooking but were also recovered in contexts that implied ritual (Maenchen-Helfen 1973). Because of the purported importance of *Amanita muscaria* in Vedic, some Iranian, and Siberian cultures (Allegro 1970; Wasson 1968, 1971) and the general importance attached to mushrooms in eastern Europe (see Eastern European Pagans), it is not surprising to find mushroom motifs in the artifacts of a people as widely traveled as the Hun nomads, but the meaning of these motifs is entirely conjectural.

Fungi in Ancient European Folklore

The relative antiquity of folklore is a subject of considerable debate, as are the extent to which and the processes by which ancient (often pagan) folkways have been altered by centuries of acculturation to Christian, Islamic, or, more recently, secular beliefs. In general, it is assumed that references to fairies, pixies, trolls, witches, and the like represent buried (or in very exceptional cases, not so buried) paganism. Folktales, legends, charms, and folk beliefs have been cataloged, often intensely, and analyzed for indications of ancient beliefs and practices. Etymological studies have been conducted with the same objective of finding a window into the remote past. Baltic and Celtic folkways in particular have been studied and analyzed because of their purported proximity to ancient or pagan habits. Folkways involving fungi are no exception, as the Baltic and Celtic folklore analyzed below attests.

An important complication in such analysis is neopaganism, the adoption of pagan (or purportedly pagan) rituals and beliefs, mainly by ex-Christian apostates. Neopaganism, regardless of its moral or spiritual values, should be excluded from the present analysis whenever possible because the focus of the present study is antiquity. Some persons who have studied the impact of fungi in antiquity have made statements in their writings or interviews indicating their neopagan orientation or personal interest in experimentation with psychoactive substances. This should not discount the evidence they present regarding the roles of fungi in the past, but the evidence should receive the same thorough scrutiny as any evidence that originates from investigators with a discernible agenda.

Documentation: Fungi in Ancient European Folklore

Eastern European Pagans

Gimbutas (1999) repeatedly links old pagan beliefs and practices in Europe with remnants in Baltic folklore: "Neither the presence of the Indo-Europeans nor the five centuries of intensive war between paganism and Christianity exterminated the oldest layer of Baltic beliefs." Of particular interest here is the association of the *Kaukai* (a category of chthonic spirits) with fungi in various contexts:

> *Many words that have the root* kauk- *relate to mushrooms:* kauka-grybis (grybis = *mushroom*) *refers to a whole group* [sic] *of mushrooms,* Phallus impudicus; kaukatiltis (tiltas = *bridge*) *represents a place where many mushrooms can be found;* kaukoratis (rata = *circle or wheel*) *is a cluster of mushrooms. There is a saying: "Don't light any fires on* kaukoratis, *because* Kaukeliai *(diminutive of* Kaūkas*) might come there at night."* (Gimbutas 1999, p. 211)

Note that *Kaukai* were manifestations of earth power in the cycle of life and death in the Baltic religion. According to Gimbutas (1999), this religion represented perhaps "the greatest repository of Old European beliefs and traditions. Here, pagan religion persisted not through millennium-old historic sagas but via oral traditions and customs that endured to the twentieth century." She adds, "The association of Kaukai with the skull and the head as the soul's abode, their association with roots and mushrooms, and their glandular, embryonic appearance leave little doubt about the essence of these chthonic beings." Gimbutas was a colorful and controversial figure whose conclusions are not always shared by her more mainstream academic colleagues (Berggren and Harrod 1996, Christ 2000, Meskell 2000). However, Biezais (2006) characterizes Baltic religion as a repository of pagan belief for the same reasons given by Gimbutas and lists the "Mother of the Mushrooms" among the forest divinities of the Balts. Many additional links between mushrooms and Slavic language or folk beliefs have been thoroughly documented (Toporov 1985).

Brown and Novick's 1992 interview with Gimbutas, posted online, specifically brings up mushrooms at one point, providing an example of how a segment of the counterculture has embraced Gimbutas's perspective. As noted in regard to the ideas of Robert Graves, C. A. P. Ruck, and others, the evidence of fungi in ancient times, including that presented by Gimbutas, needs to be assessed on its own merits and not necessarily as part of a particular view of ancient religion, patriarchy, or matriarchy. The need to separate ancient folk beliefs regarding fungi from attitudes engendered by modern popular culture has also been noted by Jürgenson (n.d.b).

At any rate, it is generally conceded that pagan beliefs have persisted in isolated parts of eastern Europe. According to *The Economist* (2005), "Mari-El and Udmurtia [two regions with Finno-Ugric populations between Moscow and the Urals] are probably the only places in Europe where shamanism (nature worship) is still an authentic, organized religion, with weddings celebrated in sacred groves." This persistence, however tenuous, of such traditions, plus the comparatively recent conquest by Christianity of the Baltic religion, make it tempting to project pagan folklore from this region (including that concerning fungi) far back into the Iron Age and earlier. Although great progress has been made in comparative studies of folklore motifs (e.g., Thompson 1955–1958; Uther 2004), consensus on the origin of folktales has not been reached, and multiple hypotheses have been advanced and debated (see Dundes 1965, Thompson 1977). It is highly plausible, but not certain, that the role of fungi in Baltic and related east European folklore is reflective of the Iron Age or before.

Some philological connections also support the antiquity of fungi in Slavic folklore. The word *baba* (whose meanings include old woman, grandmother, witch, medicine woman, midwife, female ancestor) is frequently connected with mushrooms, sometimes in a very specific context (e.g., *babka* to denote *Boletus edulis*). Stankiewicz (1958), in a study of kinship and taboo in Slavic folklore, elucidates these connections and their relation to the Slavic underworld, which he notes "is reduced to a shady, half pagan status due to the encroachments of official Christendom." Dunn (1973), in a lengthy footnote, alludes to similar parallels between Baba-Yaga (a Slavonic goddess or witch), mushrooms, and serpents. Gliwa's (2003) treatment of witches in Baltic fairy tales provides multiple examples of Baltic words that, translated loosely as witch or fairy, were linked with mushrooms, fungi, or even a specific fungus (*Merulius lacrymans*). Jürgenson (n.d.a) summarizes research on Estonian beliefs regarding the slime fungus, *Fulgio septica*, and miscellaneous other slime molds plus *Tremella* sp.

As noted already, Rubchak (1981a) interprets the pagan origins of the fire-starting ritual with mushroom tinder in Ukraine. Other examples of eastern European folk tales involving mushrooms are given by Järv (n.d.), in which a little man is held captive in a mushroom, and Warner (2002), in which demons hide under caps of toadstools during thunderstorms. Folkloristic motifs can be extraordinarily persistent; witness beliefs in "yellow as a mushroom" Baba-Yaga-like witches complete with flying mortars or broomsticks, which were widespread in nineteenth- and early twentieth-century Russia and Ukraine (Worobec 1995). Vilenskaya (n.d.) gives an experiential but also well-documented account of mushroom lore in Russia, including tales of Baba-Yaga hunting mushrooms, spirits living under mushrooms, etc. Warner (2002) summarizes the debate over whether certain folkloristic motifs in Baltic lore (none explicitly fungal) are pagan or Christian.

Western Europe

The age and origin of analogous folklore beliefs from western Europe are less certain, given the comparative remoteness of its pagan past. Some ideas seem associated with pagan roots, while others incorporate Christian elements. Common motifs included the avoidance of trespassing on fairy rings and the association of fairies with puffballs, "fairy butter," and "witches' butter" (the butters being jelly fungi, species of *Exidia* and *Tremella*). Stinkhorns (*Phallus impudicus*—the name is descriptive) and puffballs were associated with the devil. The fungus now known as *Auricularia auricula* was connected in popular superstition with Judas Iscariot and was known as "Judas's ear" or "Jew's ear" and thought to have medicinal value. The deformation of leaves and branching caused by certain fungi (primarily *Exoascus* or *Taphrina* spp.) was called "witches' broom" and was thought to be induced by flying witches. Synopses of these and other folk beliefs have been provided by Findlay (1982), Morgan (1995), and Rolfe and Rolfe (1925). Findlay also concisely recapitulates material on Greeks and Romans, modern fiction, and other aspects of fungal lore.

Common names for assorted fungi in the European folk tradition include witches' butter, troll's butter, fairy club, fairy stools, *puckfists* (fairy farts), pixie puffs, and many others (Morgan 1995). Some of these fungi ("witches' butter," "fairy butter," or "devil's butter"—all probably *Tremella* spp.) also rated mention by Thiselton-Dyer (1994 [1889]), a nineteenth-century collector of folklore pertaining to plants; he also mentioned other fungi in connection with fairies.

Although the reader is encouraged to inspect the folk beliefs from western Europe, their applicability to the time period under consideration here (Neolithic to medieval) seems more tenuous than that of the folktales and rituals from the Baltic context. For example, stories of the miscellaneous adventures of St. Peter and Christ with poisonous versus edible fungi (Morgan 1995) might devolve on pagan roots, but the Christian context is the most obvious aspect, and the same can be said of tales from Italy and elsewhere in which mushrooms are produced from St. Peter's spittle (e.g., Aarne 1961; Uther 2004). Although Wotan is a pagan Norse deity, the account by Morgan (1995) in which fly agarics are generated from his steed also has a Christian context. A Christian context is even more explicit in the several stories of the magic powers of mold from Christ's grave or the graves of churchyards (Thompson 1955–1958).

Some ethnomycologists have remarked on the connection between the fly agaric and Santa Claus, essentially deriving that connection from Santa's habit of entering and exiting houses via chimneys (a shamanistic trait), as well as the coloration of his costume and his connection with reindeer and elves (Morgan 1995). The connection between *Amanita muscaria* and chimneys is reinforced by the fact that chimney sweeps in central Europe used the fly agaric as an emblem. But again, these folkways are embedded in thoroughly Christian culture, and it is difficult to separate pagan components from Christian embellishments. Readers who find these connections far-fetched may wish, for comparison, to inspect the conjectures of Heinrich (2002) with respect to the red and white caps worn by present-day cardinals and popes, respectively. The caps are held by Heinrich to descend from analogously shaped headgear used to apply psychoactive extracts of *A. muscaria* to the tonsured pates of early Christian clergy. Perhaps the inclination of certain ethnomycologists to explain everything from Santa to the Pope by reference to *A. muscaria* is best regarded as a tribute to the human imagination.

Pictures of cute fairies and colorful mushrooms (especially *A. muscaria*) are so familiar in contemporary American and British culture that it might be assumed that this connection between fairies and the colorful and/or shapely agarics must be an ancient British tradition. Indeed, there are traditions linking the old gods to fungi in Baltic (above) and Celtic lore (below), but the cute little fairy or elf, sitting either atop or under a brightly colored mushroom, is largely a product of Victorian England. Jay (2004) provides a readable account of how fairies and little red mushrooms became part of Victorian fairy lore and how that lore was incorporated into the lore of "magic" mushrooms and the 1960s psychedelic experience.

The Celts and Fairies

Celtic folktales probably provide the best window connecting western Europe's pagan past with fungi. Although the folklore of western Europe, in contrast to that of the Baltic and some other eastern European regions, has evolved for many centuries in a Christian environment, there are especially well-documented instances of persistent pagan motifs in Celtic folklore.

One especially persistent motif associates fairies with fairy rings and mushrooms that grow in rings. Many stories relate to the dancing of fairies on or within these rings, the risks of entering such rings (especially when fairies are present), etc.: "A vast amount of legendary lore is connected with 'fairy rings'—little circles of vivid green grass frequently observed in the darker green of old pastures. . . . These fairy rings have, time out of mind, been held in great reverence by the country people [who] carefully avoid trespassing on the magic circles" (Wood-Martin 1970 [1902]). The encyclopedic treatise of Evans-Wentz (1911) contains repeated references to fairy rings from Ireland, Scotland, Wales, Cornwall, Brittany, and elsewhere, and Morgan (1995) summarizes patterns of similar tales from Wales and England. Frazer (1955b) notes that the use of iron objects as protection against fairies was part of a general pattern in which iron was regarded with superstitious aversion by pre–Iron

Age deities, spirits, and their devotees; the folklore collector Mooney (1887) stated much the same thing. Frazer's opinion reinforces the repeated references in Evans-Wentz (1911), Wood-Martin (1970 [1902]), and others in which the fairy faith is held to originate in pre-Christian belief, so presumably the beliefs regarding fairy rings also originated in pagan times.

More modern compilations also contain references to fairy rings, including allusions to fairies dancing in these rings and the mishaps that befall persons who step into the rings, e.g., a person who puts just one foot into the ring can see the pixies but still escape, but those placing both feet into the ring become prisoners (Briggs 1976). My personal favorite is in "The Fairies' Dancing-Place": Lanty M'Clusky builds a house on "one of those beautiful green circles that are supposed to be the playground of the fairies" but, even after all his labor, must pull the house down and rebuild elsewhere because the fairies are disgruntled. He is rewarded for his cooperation by the discovery of a "kam" of gold when digging the foundation for the substitute house (Yeats 1973 [1892]).

Fungi figure in Celtic folktales in other ways besides fairy rings. Wilde's (1979 [1852]) account of "The Pookaun Infants" is clearly from a Christian context, but the association of unbaptized persons with fungi implies that fungi were regarded as connected with paganism. In this story, young children inside a fairy cave are all blindfolded and are seen sitting on "pookauns" (defined in a footnote as "mushrooms, fairy-stools, or puffballs"); these children are the souls of infants that were never baptized. Morgan (1995) conveys an example from the Welsh epic the *Mabinogion* (apparently written down sometime in the thirteenth century but containing abundant pre-Christian lore), in which toadstools are transformed into magic shields. In another tale from Wales, fairy bread, when served to mortals, must be consumed the same day it is given, otherwise it will be transformed into toadstools (Baughman 1966).

The Anglo-Saxon Charms

In accord with the general notion of Anglo-Saxon culture as mycophobic (see Edible Fungi), it is difficult to find references to mushrooms in early Anglo-Saxon folklore. However, fermented beverages, leavened bread, and other manifestations of fungi (but not mushrooms) figure repeatedly in the old Anglo-Saxon charms (Grendon 1909). These charms are usually in an explicitly Christian context, but there is much evidence that their antecedents were pagan and sometimes related to analogous charms in Norse paganism (Grendon 1909). The most common manifestations of things fungal are the repeated uses of mead, ale, wine, or "yeast" (presumably batter or broth used to initiate fermentation) in verse or ritual, but there are also instances involving lichen and ringworm. The use of charms against ringworm persisted until the 1970s (Davies 1998).

Etruscan Echoes

Ancient Etruscan mythology and Italian peasant folklore of the late nineteenth century may have an enduring link. Leyland (2003 [1897]) recounted the connection between Norcia, Etruscan goddess of truffles, and Nortia, "still very generally known in La Romagna," who specialized in making midnight mushrooms. It is also possible that certain ancient Etruscan mycophagous traditions have persisted to modern times. Andrews (1958), investigating the kinds of mint used in Greek and Roman cooking, notes the purported derivation of the names for certain species and subspecies of mint from the Etruscan and the continued use of these particular mints for seasoning mushrooms (*pioppini*, the black poplar mushroom *Agrocybe aegerita*) in modern Italy.

Ideas of the Ancients on Fungal Biology

The ability of ancient peoples to understand fungal biology was understandably limited. Although they could reason carefully, a fundamental constraint was the rarity of testing hypotheses via experiments that systematically included the use of controls. Prevalence of superstition was also a factor. The most conspicuous technological limitation was simply lack of a microscope. Primitive glass lenses were made in antiquity, but there are no records of such lenses being used to construct a microscope. Without knowledge of spores or gametes, any foray into inductive reasoning and experimentation was largely incapable of explaining vast areas of biology. Prayers, charms, and sacrifices were the rule in attempts to influence the health of crops, people, or animals. Nonetheless, some ancients did ask pertinent questions regarding reproduction, contagion, and protection or therapy. Theophrastus in particular was so percipient that it has been suggested he brought pinhole apertures close to his eye when viewing small objects.

Documentation: Ideas of the Ancients on Fungal Biology

Biology without a Microscope

Historians of science have often contrasted an empirical perspective, based on sensory perception, with knowledge gained via reasoning and noted the dominance of the latter in ancient times. "If the early philosophers were inclined to favor reason over sense," however, "this tendency was neither universal nor without qualification" (Lindberg 1992). In fact, there are numerous records of meaningful speculation on fungal biology by persons in the ancient world. Buller (1915, p. 31) recounted the following comment of Pliny:

> *Amongst the most wonderful of all things is the fact that anything can spring up and live without a root. These are called Truffles (tubera); they are surrounded on all sides by earth, and are supported by no fibers, or only by hair-like root threads. . . . Now whether this imperfection of the earth (vitium terrae)—for it cannot be said to be anything else—grows, or whether it has at once assumed its full globular size, whether it lives or not, are matters which I think cannot be easily understood. In their being liable to become rotten, these things resemble wood.*

Buller also summarized and quoted Plutarch's explanation of "why truffles are thought to be produced by thunder." It was hypothesized that thunder contained a generating fluid that, when mixed with heat and piercing the earth, produced truffles. It was conceded that truffles are not plants, despite the obvious fact that they are nourished by rain. According to Buller, Theophrastus held a similar opinion, linking truffles with rain and thunder. Theophrastus also ventured the possibility that truffles were produced from seed and provided some circumstantial evidence for transport of such seed by flooding. However, Buller noted that in general, ancient peoples did not regard fungi as originating with seed, and he quotes Phanias (cited in Athenaeus, A.D. 230) that fungi "produce neither bloom nor any trace of generation by buds or seeds." As noted in Poisonous Fungi and Mycotoxins, Nicander posited the production of a fungus by "evil ferment of the earth" (Buller 1915, Houghton 1885). And as noted in Fungi Used for Medicinal Purposes and Other Technologies, Dioscorides held that *agaricum* (*Laricifomes officinalis*) was present in two sexes, male and female. At first sight, this might appear similar to opinions of modern biologists, who have documented numerous instances of sexual reproduction in fungi, but the sexes of *agaricum* in Dioscorides were probably just fruiting bodies of different appearance (Buller 1915).

The ideas of Greco-Roman authors on the production of toxins by fungi, summarized by Buller (1915) and Houghton (1885), have been presented in the section on poisonous fungi above. Houghton (1885) transmitted many of the same ancient opinions on fungal biology as Buller (1915) did. The ideas linking truffles with thunder are very briefly compared by Ainsworth (1976) with analogous ideas from Central America and southern Mexico on *Amanita muscaria* and thunder. Folk beliefs worldwide link fungi of various sorts with storms and thunder (Toporov 1985). My own favorite is the Persian legend of Mama, "who, when she shakes her baggy trousers, dispatches a swarm of flying lice to the earth, and from these lice, after a thunderstorm, an abundance of mushrooms spring up" (Suhr 1974).

Of course, ancient peoples' knowledge of biology was limited by their ignorance of microscopic life forms. In short, they had no microscopes. Although some glass and crystal lenses were manufactured in antiquity, many were merely decorative and even those putatively used to magnify were incapable of resolving microscopic organisms (Plantzos 1997). However, the observations of Theophrastus were often so meticulous that Carefoot and Sprott (1967) thought it quite possible that he might have used a tiny hole or slit in paper held close to the eyes as a sort of primitive lens. Very small objects (but not microbes) can be seen with clarity by such a method. The Roman Varro, in *On Agriculture* (1934 [first century B.C.], trans. W. E. Hooper), speculated that tiny creatures, originating from drying marshes and so small as to be invisible to the eye, might be the source of some maladies: "In the neighborhood of swamps . . . there are bred certain minute creatures which cannot be seen by the eyes, and which float in the air and enter the body through the mouth and nose and there cause serious diseases." He called these creatures *animalculae* and noted their adaptation to high humidity. This is probably as close as ancient peoples came to formulating a germ theory of disease.

Some Additional Hypotheses Regarding the Impact of Fungi in Ancient Times

The preceding sections have covered a variety of hypotheses about fungi in ancient times, but a few hypotheses merit separate examination. Those most emphasized in this section focus on the impact of plant pathogens on early agriculture. Others pertain to the impact of plant disease on the history of the Jews and the role of fermented beverages in maintaining and reinforcing social structure and status. The hypotheses pertaining to the use of fermented beverages in feasts given by chiefs or monarchs are the most buttressed with well-researched data and present a multifaceted range of reciprocal obligations, technologies, and cultures. It has been said many times that love makes the world go round, but if the hypotheses below are correct, it was more likely beer than love.

Documentation: Some Additional Hypotheses Regarding the Impact of Fungi in Ancient Times

Specific hypotheses about how fungi affected the ancient world are highly varied, comprising elements of agriculture, politics, class structure, and even evolution of consciousness. We have already examined some such conjectures; some additional hypotheses worthy of mention are summarized here.

Plant Pathogens and the History of Agriculture

A hypothesis especially far-reaching in its implications is the "honeymoon" hypothesis of Dark and Gent (2001), who proposed that pests and diseases of cereal crops were less damaging in the early agriculture of temperate northwest Europe than in medieval periods and later. The authors argued that arthropods and microbial cereal pests had evolved on crops in warmer, semiarid climates and that it took considerable time for these destructive agents to adapt to the conditions of agriculture in a cooler, temperate regime, or to be transported there: "Pests and diseases which were present in southwest Asia, southeast Europe, and the Mediterranean would not necessarily keep pace with the colonization rate of the crops." They used historical data on yields from ancient and medieval documents (admittedly sketchy), plus results of some modern experiments, to support the idea that yields were low in medieval northwest Europe, not just in comparison with modern farming, but also relative to yields obtained during and before Roman occupation. Constraints of climate and soil nutrition were judged less important than the relative impacts of pests, diseases, and weeds. Interestingly, the protective effect of hulls (glumes, spiky awns) was proposed as an additional possible explanation for resistance of early cereals to pests. "Naked" cereals were not widely grown until after the Roman period. Transport of grain during Roman times may have contributed to the spread of pests and weeds into northwest Europe. Dark and Gent (2001) also suggest that the ergot so prevalent in medieval times was ill adapted to the hulled cereals grown earlier. This "honeymoon" may have covered the Neolithic Linearbandkeramik culture of central Europe (sixth millennium B.C.) and the much later development of farming in northern and western Europe. But by the late Iron Age and Roman times, transport of seeds and foodstuffs probably introduced seedborne fungal diseases, many weedy plants, and arthropod pests. By medieval times, the Neolithic honeymoon was over, and "Paradise Lost" yields characterized agriculture until the advent of modern farming.

Independently, Carefoot and Sprott (1967) earlier proposed a similar idea when they suggested that barberry, alternate host for stem rust of wheat, was introduced into medieval Europe via Sicily and that this introduction occasioned increased crop disease and hunger. Schumann (1991) implied a much earlier introduction of barberry into Italy during Roman times.

Another set of hypotheses posits extreme impact of a single plant pathogen on cropping systems of the ancient Levant. Abbo et al. (2003) review the cultivation and characters of chickpea (*Cicer arietinum*) and its wild relative *C. reticulatum* and present a number of alternative hypotheses to explain a gap in the archaeological record for chickpea and to explain why chickpea is nearly unique among the foundation crops of the ancient Near East in being grown under a summer cropping system. After reviewing in detail the plants' agronomic traits (pod indehiscence, dormancy, vernalization insensitivity, etc.), the archaeological record, and the writings of classical scholars (Theophrastus, Pliny), the authors conclude that susceptibility to Ascochyta blight (caused by *Ascochyta rabiei*) was likely determinative of both the gap in the archaeological record and the necessity for cropping chickpea in the absence of seasonal rains. They raise the possibility that dispersal of the seedborne *A. rabiei* from the Balkans to the Levant may have truncated early winter cropping of chickpea, and that cropping of chickpea under a summer system may in turn have facilitated the introduction of warm-season sorghum, sesame, and other species into the ancient Levant. The implications of practicing both winter and summer cropping are profound with respect to crop rotation, animal husbandry, food security, and dispersal of agronomic techniques.

Carefoot and Sprott (1967) stress the impact of rust of cereal grains on human history. In a short paragraph, they suggest that data from tree rings and historical writings demonstrate a wetter than normal climatic cycle for the Mediterranean in the first three centuries A.D. and that the concomitant increase in rusts (and decrease in grain supply) was a factor in the decline of the Roman Empire. The possible role of rusts in a Biblical context was more extensively discussed. They hypothesize that the east wind that blasted the seven thin ears in the Bible did so because it brought rains and rust spores. Climatological data are adduced in support of this hypothesis. The authors consider grain to have been heavily rusted throughout the region at the time of the Biblical seven thin ears. However, grain from previous years had been stored in abundance in Egypt by the prudent Joseph, there in service of the Pharaoh. Jacob sent his sons there for grain, and later he and his clan migrated to Egypt. Thus the stage was set for the eventual enslavement of the Jews in Egypt, their escape out of bondage under Moses, the Ten Commandments, etc. Carefoot and Sprott conclude, "Once again we are forced to realize that the impact of a single plant disease can reverberate through the centuries from ancient times to the present." (Other examples they consider, outside the scope of our reference period, include ergot in the Middle Ages and the Irish potato famine.) The proportions of myth versus history in the Biblical accounts of bondage and exodus have been endlessly debated, but it seems reasonably established by Pharaonic texts that pastoralists ("Shasu" or "Shosu") who worshipped a god rendered as "Yhw" ("Yahweh") inhabited southern Transjordan and the Nile delta in the appropriate time frame (Dever 2003, Finkelstein and Silberman 2001). The basic premise that an early Hebrew tribe (perhaps Joseph's clan) migrated to Egypt in response to rust-induced grain scarcity in the southern Levant is not far-fetched. It is also reasonable to assume that rust affected grain supply in the Roman Empire, but the sudden and total loss of the North African breadbasket to the Vandals (Heather 2006) probably had more impact on the empire's decline than did the cumulative ravages of *Puccinia* spp.

Fermented Beverages and Social Structure

Perhaps the most well-founded hypotheses (perhaps one should even say theories) for the influence of fungi on ancient peoples involve the impact of fermentation products on social structure. Beer and wine were integral to feasting, and feasting was integral to maintenance of political power and social status throughout the ancient Near East, the Mediterranean, and Europe. Both the feast itself and its organization were political and social events that reflected and reinforced the power of the organizers (Jennings et al. 2005).

Bottéro (2004) lists the provisions for a 10-day feast given by the Assyrian king Aššurnasirpal. Even allowing for exaggeration, the scale is impressive, not only for the food (1,000 barley-fed oxen; 1,000 young cattle; 1,000 fattened sheep, etc.), but also for the lavish drink (10,000 measures of beer; 10,000 skins of wine). Other kings in Mesopotamia gave similarly lavish banquets, and in Mesopotamian mythology the gods themselves dined in such style at the investiture of the high deity, Marduk. Smith (2003) remarks of Egyptian feasts from the time of the New Kingdom, "Thousands of wine amphorae from the king's jubilee celebrations were discovered . . . in trash deposits. . . . Wine apparently flowed so freely that . . . servants appear to have lopped off the tops of the jars rather than taking the time to open the seals."

As Bottéro (2004) notes, "the sharing of meals could also intervene, quite powerfully, in social and political life." Jennings et al. (2005) call such ancient feasts "important arenas of political action" and summarize the archaeological literature bearing on the social, economic, and political ramifications of feasting. They go into more detail in analyzing the processes for producing the copious quantities of fermented beverages called for by such feasts. From the perspective of ancient societies, these processes made extensive demands on agricultural supplies, pottery, storage facilities, human labor, and animal transport. Of particular interest are the processes for barley and emmer wheat beer and grape wine in the ancient Near East. The very limited "shelf life" of ancient Near Eastern beer was a prime determinant of how feast preparations were organized. Wine could be better preserved (especially if amended with suitable resins; see McGovern 2003), but the production and assembling of large quantities was still an organizational challenge.

According to Jennings et al. (2005), "wine was a symbol of power and status in these societies and tended to be used in feasts that signaled social differences." Pollock (2003) provides analogous theoretical analyses for Mesopotamian states, where the presentation and consumption of food and drink was "in the service of political, religious, and other social goals." She remarks on the significance of the representation of drink in feasting scenes (dominating the representation of food) and suggests that symbols of drinking are symbolic of commensal occasions. Very similar ideas attend analyses of feasting in the Bronze Age Aegean. Palaima (2004a,b) writes of feasting in Mycenaean Thebes, Pylos, and Knossos, "Commensal ceremonies are meant to unite communities and reinforce power hierarchies." Linear B tablets provide plentiful references to beer and wine assembled for feasts, as well as associated food dishes (Palaima 2004a,b).

The duration, importance, and function of such practices are further highlighted by Pollington's (2003) treatment of mead and feasting among the Anglo-Saxons, whose culture and history straddled the boundary between pagan and Christian. Pollington illustrates the significance of mead and beer in Anglo-Saxon culture with selections from *Beowulf* describing the great mead hall Heorot, the protocol for entering the hall, dispensations of treasures to the guests, and samples of Anglo-Saxon balladry. Mead also figures in feasting, hospitality, and balladry in the Icelandic-Norse Volsungs and Eddas and in the Germanic Nibelungenlied.

Psychoactive Fungi and the Evolution of Consciousness

Although there is evidence for the use of psychoactive fungi in ancient times, few have claimed that such use was critical for the formation of modern human consciousness. However, such claims have been advanced by McKenna (1988): "The hidden factor in the evolution of human beings, the factor that called human consciousness forth from a bipedal ape with binocular vision, involved a feedback loop with plant hallucinogens." Essentially, the argument is that early hominids became habituated to the ingestion of hallucinogenic mushrooms that bestowed an ecstatic experience, and "human language arose out of the synergy of primate organizational potential by plant hallucinogens." The evidence given in support of the hypothesis consisted largely of analogies with shamanistic practices, some unorthodox psychological theories, and some fungal biogeography and ecology (distribution of psychoactive fungal species, dependence of some fungi on herbivore dung, etc.). The primary evidence was the figures of shamans and mushrooms from the Sahara that may represent the earliest portrayal not only of mushrooms but also of shamans (McKenna 1988, Samorini 1992). To his credit, McKenna did term his ideas "speculations."

McKenna (1992) reiterated the hypothesis, provided an alternative suggestion for the identity of soma, and made extensive commentaries on drug use, ancient and modern. That experiences with psychoactive fungi may have played a role in the evolution of consciousness is not inherently illogical, but McKenna's use of phrases such as "gnosis of the vegetable mind" and "Gaian collectivity of organic life" did little to advance the hypothesis. Like Graves (1960), McKenna indicated devotion to the concept (and perhaps the literal existence) of the Great Goddess, and some of the ideas presented are not amenable to hypothesis testing in any conventional manner. McKenna's notion (expressed in a 1992 interview by Novick and Brown [n.d. online]) that psilocybin mushrooms represent an extraterrestrial form of intelligence is unlikely to gain much acceptance from mycologists who have studied and characterized the genus *Psilocybe* in terms of morphology, mating types, DNA sequence, and other characters pertinent to the study of its terrestrial evolution (e.g., Boekhout et al. 2002). His concept that ingestion of the mushrooms is the only way to test his hypothesis of extraterrestrial intelligence (Novick and Brown 1992 [n.d. online]) means that findings obtained from such testing will be excluded from conventional peer-reviewed journals. Readers in the United States wishing to indulge in McKenna's method of independent inquiry are advised to first consult Code of Federal Regulations 21CFR1308.11, Drug Enforcement Administration, Department of Justice.

A more restrained hypothesis (no extraterrestrial spores) detailed in Lewis-Williams and Pearce (2005) is specifically tailored to the Neolithic. In this model, certain trancelike or dreamlike mental states are upheld as the basis for religious concepts common to Neolithic cultures. This hypothesis is not centered on psychoactive fungi, although there are repeated references to the use of psychoactive plants or fungi. The trancelike states need not have been induced by ingestion of any psychoactive substances; they could have been achieved by other methods or circumstances such as fasting, concentration, exhaustion, or dreams. Far from being a critical element in the hypothesis, fungi are mentioned as one of multiple alternative mechanisms by which these altered mental states might have been attained. The most persuasive evidence adduced for this hypothesis was ethnographic, with emphasis on shamanistic lore and myth, plus petroglyphs and artifacts from early Neolithic settlements.

Manna from Heaven

Pegler (2002) notes the good match between the properties of manna as described in Exodus and Numbers and the characteristics of desert truffles, species of *Terfezia* and *Tirmania*. Both manna and desert truffles appear after dew, do not keep but become rotten with maggots, can be preserved by cooking, and wither under the force of the sun (Pegler 2002). Alternative hypotheses for manna include the lichen *Lecanora esculenta* (Trease and Evans 1978) in addition to many others, e.g., honeydew from insects (Moldenke and Moldenke 1952).

Conclusions

Fungi had an extraordinarily strong and varied impact on the development of ancient history. Although one can conceive of ancient peoples without leavened bread (some peoples had grain porridge regularly instead of bread), there can be little doubt that ancient social and political relations were dependent on periodic, ethanol-laced feasts and banquets. Moreover, there is evidence that in ancient times, as in more modern eras, people were occasionally displaced by phytopathogen-induced food shortages or silently poisoned by mycotoxins in their food stores. Ancient literature is replete with concerns over plant diseases, and the inability of modern scholarship to precisely diagnose these diseases from a distance of multiple centuries should not dissuade us of their importance. And although much that has been written is conjectural, there is nothing inherently implausible about, for example, Carefoot and Sprott's (1967) hypotheses on the possible role of cereal rust in Biblical events or the hypotheses of Abbo et al. (2003) on the impact of plant disease on cropping systems of the Levant.

The experiments of Hill et al. (1983), Lacey (1972), and others are evidence that ancestral Europeans probably consumed grain contaminated with mycotoxins. And as described above, there is a persuasive argument that the course of the Peloponnesian War (one of the most important civil wars in Western civilization) was influenced by mass mycotoxicosis, the plague of Athens. Some other claims that have been made for the effects of mycotoxins, such as Schoental's (1980, 1984, 1987) interpretations of Biblical history, are highly suspect. Schoental's (1991) vision of fungal spores turning an ancient civilization into homosexuals might make a good late night B movie, but it is a bad model for scientific historiography.

In ancient times, as now, moldy rot limited the useful life of clothing and structures, and human health was undoubtedly adversely affected by fungi in moist habitations. For the unlucky, persistent fungal skin infections could mean disfigurement, banishment, or worse, especially because fungal infections could be confused with Hansen's disease, leprosy. This confusion undoubtedly shattered the lives of thousands of people and occasionally entire communities.

Did fungi also play a substantial role in religious experience and ritual? As producers of ethanol (by *Saccharomyces cerevisiae* especially), most certainly. As producers of other psychoactive compounds ("entheogens"), well . . . perhaps. There is certainly abundant circumstantial evidence (often philological) to indicate some kind of role for entheogenic fungi in Indo-European (including Greco-Roman) and west Asian societies. The claim of some enthusiasts that the impact of *Amanita muscaria*, *Claviceps* species, and other psychoactive fungi was foundational and pervasive may represent a projection of the personal histories of the proponents rather than historical reality. Although several of these authors were highly competent scholars, scientists in the cultural mainstream are skeptical of the findings of colleagues who incorporate personal psychoactive experience into theory. The beautifully illustrated findings of a self-confessed toad licker (Morgan 1995: "my pupils dilated and my visual perception was enhanced by an intensity of color") are probably the most enjoyable example of works likely to be regarded as on the fringe, although a close second are the works of Robert Graves, especially Graves (1958). Graves's statement that "the fly-amanite may have changed the course of European history, and even been the reason why the common dialect of Britain and the States is not a barbarous dialect of Greek" seems (intentionally?) highly reminiscent of the famous statement of Edward Gibbon: "Perhaps the interpretation of the Koran would now be taught in the schools of Oxford, and her pulpits might demonstrate to a circumcised people the sanctity and truth of the revelations of Mohammed." However, the battle of Poitiers belongs to the realm of history, whereas the forays of Graves's *Amanita*-intoxicated warriors were constructed from the realm of myth and from Graves's undeniably fertile imagination. Graves (1958) is so replete with poetic energies that unwary ex-hippies may experience flashbacks anywhere in the pages between the Maenads and the bemushroomed "staid deaconess, . . . giggling helplessly, stark naked, on the floor."

The works of these exuberant, experientially oriented authors are entertaining, but readers must be on their guard to distinguish mycohistory from myco-fiction. Perhaps the most reasonable approach is that of Lewis-Williams and Pearce (2005), who content themselves with noting that certain fungi are among several agents and conditions potentially contributing to trancelike or dreamlike states integral to the Neolithic religious mind and its Bronze Age, Iron Age, or medieval post sequiturs. Unfortunately for ancient farmers and healers, such states of mind (whatever their metaphysical merits) were insufficiently reductionist for scientific hypothesis testing, and so the nature of fungi and their role in disease were not truly appreciated until the modern era.

If the researches of the more experientially inclined enthusiasts fail to fully persuade, the folklorists stand on firmer ground. The pan-European linkage between fungi (mostly mushrooms, jelly fungi, or puffballs) and non-Christian entities such as witches, trolls, and fairies testifies to the significance of fungi in the distant, pagan past. Moreover, the regularity with which fungi were used for tinder, dyes, and medicinal purposes, as attested not just by folklore but also by archaeology, speaks for the familiarity of European pagans with fungi. Unfortunately, most ancient folkways can only be dimly perceived through successive layers of Christian and more recent additions and embellishments. We will probably never know many specifics of how preliterate European peoples viewed and used macrofungi in pagan times. Although it is tempting to posit a cultural connection between the mushroom stones of the Neolithic Balkans and the mycophilic folkways of later eastern Europe, we will likely never know. As the mushroom cauldrons of the Huns demonstrate, mushroom imagery was literally all over the map of Europe and west Asia, but seldom within clearly defined contexts. Debate continues over the degree of ethnic and cultural continuity in Neolithic Europe, versus migration and invasion (see the introduction). What can be said with assurance is that the abundant folklore and myth relevant to fungi have roots (however ill defined) in remote antiquity. Root words in many languages, especially those of the Nostratic family (from the early Neolithic), convey intimately related thought processes linking mushrooms with descriptors for sex and sexuality, kinship, cooking, and food (Toporov 1985).[7]

The areas that are most apt to yield substantial new discoveries in coming years are studies of fungi naturally cryopreserved in glacial ice or permafrost and research on coevolution of fungi with species and landraces of crop plants indigenous to core areas of the Neolithic such as the Fertile Crescent and adjacent regions. These meldings of ancient and modern promise to be especially fruitful with regard to rates of microbial evolution vis-à-vis human agriculture and may yield data of practical utility in plant protection. Plant germplasm collections contain many of the species and landraces pertinent to the beginnings of agriculture, and culture collections contain numerous fungal plant pathogens associated with these plants. One can hope for a coalescence of data from collection-based research, studies of naturally cryopreserved microbes, and field studies conducted in centers of crop diversity, the results from which will greatly extend knowledge of the rates and directions of plant-pathogen interactions subsequent to the emergence of agriculture.

---

[7] Nostratic and other linguistic macrofamilies are still much debated. See Renfrew (2000) for the coalescence among linguistic, molecular-genetic, and archaeological approaches to prehistory. Balter (2004) supplies an analogous summary analysis, including the support from linguistics for Gimbutas's Kurgan theory.

Host-pathogen interactions are the bread and butter of plant pathology, but much of the present review has touched on subject areas that are exotic or arcane to the majority of mycologists and plant pathologists. Most of these professionals may find comfort in some basic conclusions: Despite the hype and notoriety surrounding soma and religious mysteries, most people in antiquity were probably directly affected far more by plant diseases, food spoilage, bouts of ringworm or yeast infection, and especially the uses of yeasts as leavening and fermenting agents. For then, as now, most people's lives were not spent in revelatory trances or divine ecstasies. In times when most populations were rural and poor, people tended their flocks, gardens, and fields; fought against diseases (of plants, humans, and animals) with the medicines, spells, and rituals at their disposal; and relaxed when they could with their daily bread and a sip of beer.

Literature Cited

Aarne, A. 1961. The Types of the Folktale: A Classification and Bibliography, translated and enlarged by Stith Thompson. Suomalainen Tiedeakatemia, Helsinki.

Aaronson, S. 1989. Fungal parasites of grasses and cereals: Their role as food or medicine, now and in the past. Antiquity 63:247–257.

Abbo, S., D. Shtienberg, J. Lichtenzveig, S. Lev-Yadun, and A. Gopher. 2003. The chickpea, summer cropping, and a new model for pulse domestication in the ancient Near East. The Quarterly Review of Biology 78(4):435–448.

Ackerman, R. 1975. Frazer on myth and ritual. Journal of the History of Ideas 36:115–134.

Agouridis, C. 1997. Sea routes and navigation in the third millennium Aegean. Oxford Journal of Archaeology 16(1):1–24.

Ainsworth, G. C. 1976. Introduction to the History of Mycology. Cambridge University Press, Cambridge.

Ainsworth, G. C. 1981. Introduction to the History of Plant Pathology. Cambridge University Press, Cambridge.

Ainsworth, G. C. 1986. Introduction to the History of Medical and Veterinary Mycology. Cambridge University Press, Cambridge.

Ainsworth, G. C. 1993a. Fungus infections (mycoses). Pages 730–736 in: The Cambridge World History of Human Disease, ed. K. F. Kiple. Cambridge University Press, Cambridge.

Ainsworth, G. C. 1993b. Fungus poisoning. Pages 736–738 in: The Cambridge World History of Human Disease, ed. K. F. Kiple. Cambridge University Press, Cambridge.

Ajello, L. 1974. Natural history of the dermatophytes and related fungi. Mycopathologia et Mycologia Applicata 53:93–110.

Alexopolous, C. J. 1962. Introductory Mycology, 2nd ed. John Wiley, New York.

Allegro, J. M. 1970. The Sacred Mushroom and the Cross. Bantam Books, New York.

Andrews, A. C. 1958. The mints of the Greeks and Romans and their condimentary uses. Osiris 13:127–149.

Andrews, T. 2000. Nectar & Ambrosia: An Encyclopedia of Food in World Mythology. ABC-CLIO, Santa Barbara, CA.

Armelagos, G. J., and K. Barnes. 1999. The evolution of human disease and the rise of allergy: Epidemiological transitions. Medical Anthropology 18:187–213.

Armelagos, G. J., and K. N. Harper. 2005a. Genomics and the origins of agriculture, Part one. Evolutionary Anthropology 14:68–77.

Armelagos, G. J., and K. N. Harper. 2005b. Genomics and the origins of agriculture, Part two. Evolutionary Anthropology 14:109–121.

Baker, K. F. 1965. A plant pathogen views history. Pages 36–70 in: History of Botany: Two Papers Presented at a Symposium Held at the William Andrews Clark Memorial Library, December 7, 1963. The Clark Memorial Library, University of California, Los Angeles, and The Hunt Botanical Library, Carnegie Institute of Technology, Pittsburgh.

Balter, M. 2004. Search for the Indo-Europeans. Science 303:1323–1326.

Banke, S., A. Peschon, and B. A. McDonald. 2004. Phylogenetic analysis of globally distributed *Mycosphaerella graminicola* populations based on three DNA sequence loci. Fungal Genetics and Biology 41:226–238.

Bar-Yosef, O. 2002. The Upper Paleolithic revolution. Annual Review of Anthropology 31:363–393.

Battey, N. H. 2003a. Plant culture: Thirteen seasonal pieces; Introduction—Plenty has made us poor. Journal of Experimental Botany 53:2289–2291.

Battey, N. H. 2003b. Plant culture: Thirteen seasonal pieces; February—Constructing a corymb. Journal of Experimental Botany 54:605–608.

Battilani, P., and A. Pietri. 2002. Ochratoxin A in grapes and wine. European Journal of Plant Pathology 108:639–643.

Baughman, E. W. 1966. Type and Motif-Index of the Folktales of England and North America. Indiana University Folklore Series, no. 20. Mouton & Co., The Hague.

Beard, M. 1992. Frazer, Leach, and Virgil: The popularity (and unpopularity) of the Golden Bough. Comparative Studies in Society and History 34:203–224.

Bednarik, R. C. 2003. Seafaring in the Pleistocene. Cambridge Archaeological Journal 13(1):41–66.

Belfiore, E. 1986. Wine and catharsis of the emotions in Plato's laws. The Classical Quarterly, New Series 36:421–427.

Bellard, C. G. 1995. The first colonization of Ibiza and Formentera (Balearic Islands, Spain): Some more islands out of the stream? World Archaeology 26(3):442–455.

Bellemore, J., I. M. Plant, and L. M. Cunningham. 1994. Plague of Athens—fungal poison? Journal of the History of Medicine and Allied Sciences 49:521–545.

Bellwood, P. 2005. First Farmers: The Origins of Agricultural Societies. Blackwell, Oxford.

Benjamin, D. R. 1995. Mushrooms: Poisons and Panaceas. W. H. Freeman and Co., New York.

Bennett, J. W., and M. Klich. 2003. Mycotoxins. Clinical Microbiology Reviews 16:497–516.

Berger, M. A. 1992. Predynastic animal-headed boats from Hierakonpolis and southern Egypt. Pages 107–120 in: The Followers of Horus: Studies Dedicated to Michael Allen Hoffman 1944–1990, ed. R. Friedman and B. Adams. Egyptian Studies Association Publication no. 2, Oxbow Monograph 20. Oxbow Books, Oxford.

Berggren, K., and J. B. Harrod. 1996. Understanding Marija Gimbutas. Journal of Prehistoric Religions 10:70–73.

Bernicchia, A., M. F. Fugazzola, V. Gemelli, B. Mantovani, A. Lucchetti, M. Cesari, and E. Speroni. 2006. DNA recovered and sequenced from an almost 7000 y-old Neolithic polypore, Daedaleopsis tricolor. Mycological Research 110:14–17.

Bernstein, L. 2003. A historical perspective of occupational asthma with a special account of the contribution by Professor Jack Pepys. American Journal of Respiratory and Critical Care Medicine 167:451–452.

Bersu, G. 1940. Excavations at Little Woodbury, Wiltshire. Part I. The settlement as revealed by excavation. Proceedings of the Prehistoric Society, New Series 6:30–111.

Betancourt, P. P. 2003. The impact of Cycladic settlers on early Minoan Crete. Mediterranean Archaeology and Archaeometry 3:3–12.

Biezais, H. 2006. Baltic religion. Encyclopedia Britannica Online, www.search.eb.com/eb/article-65448.

Blanchette, R. A. 2000. A review of microbial deterioration found in archaeological wood from different environments. International Biodeterioration & Biodegradation 46:189–204.

Blatner, K.A. 2000. Special forest product markets in the Pacific Northwest with global implications. Pages 42–53 in: Special Forest Products: Biodiversity Meets the Marketplace, ed. N. C. Vance and J. Thomas. Pacific Northwest Research Station, USDA Forest Service, General Technical Report GTR-WO-63.

Boekhout, T., J. Stalpers, S. J. W. Verduin, J. Rademaker, and M. E. Noordeloos. 2002. Experimental taxonomic studies in Psilocybe sect. Psilocybe. Mycological Research 106:1251–1261.

Bonman, J. M., H. E. Bockelman, L. F. Jackson, and B. J. Steffenson. 2005. Disease and insect resistance in cultivated barley accessions from the USDA National Small Grains Collection. Crop Science 45:1271–1280.

Bonman, J. M., H. E. Bockelman, B. J. Goates, D. E. Obert, P. E. McGuire, C. O. Qualset, and R. J. Hijmans. 2006. Geographic distribution of common and dwarf bunt resistance in landraces of Triticum aestivum subsp. aestivum. Crop Science 46:1622–1629.

Bonser, W. 1963. The Medical Background of Anglo-Saxon England: A Study in History, Psychology and Folklore. The Wellcome Historical Medical Library, London.

Borgen, A. 2000. Hvedens stinbrand—en udfordring for principperne for økologisk jordburg. Ph.D. thesis, Institut for Jordbrugsvidenskab, Den Kongelige Veterinær- og Landbohøjskole, http://orgprints.org/263/01/Ph.D.-afhandling.pdf.

Bottéro, J. 2004. The Oldest Cuisine in the World: Cooking in Mesopotamia. University of Chicago Press, Chicago.

Braidwood, R. J., J. D. Sauer, H. Helbaek, P. C. Mangelsdorf, H. C. Cutler, C. S. Coon, R. Linton, J. Stewart, and A. L. Oppenheim. 1953. Symposium: Did man live by beer alone? American Anthropologist, New Series 55:515–526.

Braun, T. 1995. Barley cakes and emmer bread. Pages 25–37 in: Food in Antiquity, ed. J. Wilkins, D. Harvey, and M. Dobson. University of Exeter Press, Exeter, U.K.

Breatann, G. m. R. d. n.d. *Panis Romanum:* The history and evolution of bread in Rome, rev., webzoom.freewebs.com/earlyperiod/PANIS%20ROMANVM.pdf.

Brenton, B. P. 1990. Mycotoxin analysis of stored barley from Iron-Age-type storage pits: Implications for human health and nutrition (Abstr.) American Journal of Physical Anthropology 81:199.

Briffault, R. 1931. The Mothers: The Matriarchal Theory of Social Origins. McMillan, New York.

Briggs, K. 1976. An Encyclopedia of Fairies, Hobgoblins, Brownies, Bogies, and Other Supernatural Creatures. Pantheon, New York.

Broodbank, C. 1993. Ulysses without sails: Trade, distance, knowledge and power in the early Cyclades. World Archaeology 24(3):315–331.

Broodbank, C., and T. F. Strasser. 1991. Migrant farmers and the Neolithic colonization of Crete. Antiquity 65:233–245.

Brothwell, D., and P. Brothwell. 1969. Food in Antiquity: A Survey of the Diet of Early Peoples. Praeger, New York.

Brown, D. J., and R. M. Novick. 1992 (n.d. online). Learning the language of the goddess, with Marija Gimbutas. In Mavericks of the Mind, Internet ed., www.levity.com/mavericks.gim-int.htm.

Brown, T. A. 1999. How ancient DNA may help in understanding the origin and spread of agriculture. Philosophical Transactions of the Royal Society of London B 354:89–98.

Buckland, P. C. 1981. The early dispersal of insect pests of stored products as indicated by archaeological records. Journal of Stored Products Research 17:1–12.

Buller, A. H. R. 1915. The fungus lore of the Greeks and Romans. Transactions of the British Mycological Society 5:21–66.

Burns, T. S. 2003. Rome and the Barbarians, 100 B.C.–A.D. 400. Johns Hopkins University Press, Baltimore.

Burres, C. L. 1995. Preserved hair and associated keratinophilous fungi from late Pleistocene scat. Journal of Vertebrate Paleontology 15(3 Supplement):21A.

Buurman, J., B. van Geel, and G. B. A. van Reenen. 1995. Paleoecological investigations of a Late Bronze Age watering-place at Bovenkarspel, The Netherlands. Pages 249–270 in: Neogene and Quaternary Geology of North-West

Europe, ed. G. F. W. Herngreen and L. van der Valk. Mededelingen Rijks Geologische Dienst 52.

Cahill, N. 2002. Household and City Organization at Olynthus. Yale University Press, New Haven.

Callender, M. H. 1965. Roman Amphorae. Oxford University Press, London.

Caplice, R. 1974. An apotropaion against fungus. Journal of Near Eastern Studies 33:345–349.

Carcopino, J. 1940. Daily Life in Ancient Rome. Yale University Press, New Haven.

Carefoot, G. L., and E. R. Sprott. 1967. Famine on the Wind: Man's Battle Against Plant Disease. Rand McNally & Co., Chicago.

Carter, R. 2006. Boat remains and maritime trade in the Persian Gulf during the sixth and fifth millennia BC. Antiquity 80:52–63.

Castoria, R., G. Lima, R. Ferracane, and A. Ritieni. 2005. Occurrence of mycotoxin in farro samples from southern Italy. Journal of Food Protection 68:416–420.

Cato, Marcus Porcius. 1934. On Agriculture, trans. W. D. Hooper. Loeb Classical Library, Harvard University Press, Cambridge, MA.

Cavalieri, D., P. E. McGovern, D. L. Hartl, R. Mortimer, and M. Polsinelli. 2003. Evidence for S. cerevisiae fermentation in ancient wine. Journal of Molecular Evolution 57:S226–S232.

Chaddick, P. R., and F. F. Leek. 1972. Further specimens of stored product insects found in ancient Egyptian tombs. Journal of Stored Products Research 8:83–86.

Chartrain, L., P. A. Brading, and J. K. M. Brown. 2005. Presence of the *Stb6* gene for resistance to septoria tritici blotch (*Mycosphaerella graminicola*) in cultivars used in wheat-breeding programmes worldwide. Plant Pathology 54:134–143.

Chernoff, M. C., and S. M. Paley. 1998. Dynamics of cereal production at Tell el Ifshar, Israel during the Middle Bronze Age. Journal of Field Archaeology 25:397–416.

Chester, K. S. 1946. The nature and prevention of the cereal rusts as exemplified in the leaf rust of wheat. Chronica Botanica, Waltham, MA.

Christ, C. P. 2000. Reading Maria Gimbutas. NWSA Journal 12:169–173.

Christakis, K. S. 1999. Pithoi and food storage in Neopalatial Crete: A domestic perspective. World Archaeology 31:1–20.

Christensen, C. M. 1973. Loss of viability in storage: Microflora. Seed Science & Technology 1:547–562.

Ciferri, O., P. Tiano, and G. Mastromei (eds.). 2000. Of Microbes and Art: The Role of Microbial Communities in the Degradation and Protection of Cultural Heritage. Kluwer Academic/Plenum Publishers, Dordrecht, The Netherlands.

Cilliers, L., and F. P. Retief. 2000. Poisons, poisoning and the drug trade in ancient Rome. Akroterion 45:88–100.

Clark, S. H., and K. J. Edwards. 2004. Elm bark beetle in Holocene peat deposits and the northwest European elm decline. Journal of Quaternary Science 19:525–528.

Clement, S. L., L. R. Elberson, N. A. Bosque-Pérez, and D. J. Schotzko. 2004. Detrimental and neutral effects of wild-barley-*Neotyphodium* fungal endophyte associations on insect survival. Entomologia Experimentalis et Applicata 114:119–125.

Colledge, S., J. Conolly, and S. Shennan. 2004. Archaeological evidence for the spread of farming in the eastern Mediterranean. Current Anthropology 45(Supp.):S35–S58.

Columella, Lucius Junius Moderatus. 1941a. De Re Rustica I, trans. H. B. Ash. Loeb Classical Library, Harvard University Press, Cambridge, MA.

Columella, Lucius Junius Moderatus. 1941b. De Re Rustica II, trans. E. S. Forster and E. H. Heffner. Loeb Classical Library, Harvard University Press, Cambridge, MA.

Comandin, O., and A. C. Rinaldi. 2004. Tracing megafaunal extinctions with dung fungal spores. Mycologist 18(4):140–142.

Cooper, A., and H. K. Poinar. 2000. Ancient DNA: Do it right or not at all. Science 289:1139.

Cordier, F. S. 1870. Les champignons de la France. J. Rothschild, Paris.

Corner, E. J. H. 1950. Report on the fungus-brackets from Star Carr, Seamer. Proceedings of the Prehistoric Society 16:123–124.

Coughanowr, E. 1979. The meaning of μολοβρός in Homer. The Classical Quarterly, New Series 29:229–230.

Craig, O. E., J. Chapman, C. Heron, L. H. Willis, L. Bartosiewicz, G. Taylor, A. Whittle, and M. Collins. 2005. Did the first farmers of central and eastern Europe produce dairy foods? Antiquity 79:882–894.

Cruse, A. 2004. Roman Medicine. Tempus, Brimscombe Port, Gloucestershire, U.K.

Currid, J. D., and A. Navon. 1989. Iron Age pits and the Lahav (Tell Halif) grain storage project. Bulletin of the American Schools of Oriental Research 273:67–78.

Curvers, H. H., and G. M. Schwartz. 1990. Excavations at Tell al-Raqa'i: A small rural site of early urban northern Mesopotamia. American Journal of Archaeology 94:3–23.

Czembor, J. H. 2001. Sources of resistance to powdery mildew (*Blumeria graminis* f. sp. *hordei*) in Moroccan barley land races. Canadian Journal of Plant Pathology 23:260–269.

Czembor, J. H. 2002. Resistance to powdery mildew in selections from Moroccan barley landraces. Euphytica 125:397–409.

Dalby, A. 1996. Siren Feasts: A History of Food and Gastronomy in Greece. Routledge, London.

Dalby, A. 2004. Bacchus: A Biography. Getty Publications, Los Angeles.

Darby, W. J., P. Ghalioungui, and L. Grivetti. 1977. Food: The Gift of Osiris, Vol. 2. Academic Press, London.

Dark, P., and H. Gent. 2001. Pests and diseases of prehistoric crops: A yield "honeymoon" for early grain crops in Europe? Oxford Journal of Archaeology 20(1):59–78.

Davies, O. 1998. Charmers and charming in England and Wales from the eighteenth to the twentieth century. Folklore 109:41–52.

Davis, J. L. 1992. Review of Aegean prehistory I: The islands of the Aegean. American Journal of Archaeology 96:699–756.

Dever, W. G. 2003. Who Were the Early Israelites and Where Did They Come From? Wm. B. Eerdmans, Grand Rapids, MI.

Dineley, M., and G. Dinely. 2000. Neolithic ale: Barley as a source of malt sugars for fermentation. Pages 137–153 in: Plants in Neolithic Britain and Beyond. Neolithic Studies Group Seminar Papers 5, ed. A. S. Fairbairn. Oxbow Books, Oxford.

Dufour, L. 1891. Atlas des champignons comestibles et vénéneux. Librarie des sciences naturelles, Paris.

Dugan, F. M. 2006. The Identification of Fungi: An Illustrated Introduction with Keys, Glossary, and Guide to Literature. American Phytopathological Society, St. Paul, MN.

Dugan, F. M. Fungi, folkways and fairy tales: Mushrooms and mildews in stories, remedies and rituals, from Oberon to the Internet. North American Fungi 3, www.pnwfungi.org. (In press)

Dundes, A. 1965. The Study of Folklore. Prentice-Hall, Englewood Cliffs, NJ.

Dunn, E. 1973. Russian use of *Amanita muscaria: A* footnote to Wasson's soma. Current Anthropology 14:488–492.

Economist. December 24, 2005. Finno-Ugrics: The dying fish swims in water. Vol. 377(8458):73–74.

Edens, C. 1992. Dynamics of trade in the ancient Mesopotamian "world system." American Anthropologist (New Series) 94:118–139.

Eliade, M. 1972. Zalmoxis: The Vanishing God; Comparative Studies in Religions and Folklore of Dacia and Eastern Europe. University of Chicago Press, Chicago.

Elmholt, S. 2004. Susceptibility of spelt to ochratoxin A producing fungi. DARCOFenews, www.darcof.dk/enews/june04/toxin.html.

Elmholt, S., and P. H. Rasmussen. 2005. *Penicillium verrucosum* occurrence and ochratoxin. A contents in organically cultivated grain with special reference to ancient wheat types and drying practice. Mycopathologia 159:421–432.

Eluère, C. 1998. The golden treasures of the European Bronze Age. Pages 168–171 in: Gods and Heroes of the European Bronze Age, ed. K. Demakopoulou, C. Eluère, J. Jensen, A. Jockenhövel, and J.-P. Mohen. Thames and Hudson, London.

Encyclopedia Britannica Online. 2006. Frazer, Sir James George. www.search .eb.com/eb/article-9035218.

Engler, A. 1897. De Natürlichen Planzenfamilien, I Teil. Englemann, Leipzig.

Evans-Wentz, Y. W. 1911. The Fairy Faith in Celtic Countries. Oxford University Press, Oxford (reprint, with foreword by K. Raine, Colin Smythe, London, 1977).

Faas, P. 2003. Around the Roman Table: Food and Feasting in Ancient Rome. Palgrave MacMillan, New York.

Fabing, H. D. 1956. On going berserk: A neurochemical inquiry. Scientific Monthly 83:232–237.

Fall, P. L., L. Lines, and S. E. Falconer. 1998. Seeds of civilization: Bronze Age rural economy and ecology in the southern Levant. Annals of the Association of American Geographers 88(1):107–125.

Fay, J. C., and J. A. Benavides. 2005. Evidence for domesticated and wild populations of *Saccharomyces cerevisiae*. PLoS Genetics 1(1):66–71; www.plosgenetics.org.

Fern, K. n.d. Plants for a Future: Edible, medicinal and useful plants for a healthier world. www.pfaf.org/index.html.

Fernández-Armesto, F. 2002. Near a Thousand Tables: A History of Food. Simon & Schuster, New York.

Ferreira, E. S. B., A. N. Hulme, H. McNab, and A. Quye. 2004. The natural constituents of historical textile dyes. Chemical Society Reviews, DOI: 10.1039/b305697j, http://pubs.rsc.org/ej/CS/2004/b305697j/.

Fetch, T. G., Jr., B. J. Steffenson, and E. Nevo. 2003. Diversity and sources of multiple disease resistance in *Hordeum spontaneum*. Plant Disease 87:1439–1448.

Findlay, W. P. K. 1982. Fungi: Folklore, Fiction, and Fact. Richmond Publ., Richmond, Surrey, U.K.

Finkelstein, I., and N. A. Silberman. 2001. The Bible Unearthed: Archaeology's New Vision of Ancient Israel and the Origin of Its Sacred Texts. Simon & Schuster, New York.

Fischer, C. 1987. Review of Lindow Man: The Body in the Bog. The Quarterly Review of Biology 62:469–470.

Flint-Hamilton, K. B. 1999. Legumes in ancient Greece and Rome: Food, medicine, or poison? Hesperia 68:371–385.

Florian, M.-L. 1997. Heritage Eaters: Insects and Fungi in Heritage Collections. James & James, London.

Forbes, H., and L. Foxhall. 1995. Ethnoarchaeology and storage in the ancient Mediterranean: Beyond risk and survival. Pages 69–86 in: Food in Antiquity, ed. J. Wilkins, D. Harvey, and M. Dobson. University of Exeter Press, Exeter, U.K.

Frazer, J. G. 1898. Pausania's description of Greece. (Trans.) Vol. I, p. 94. Macmillan and Co., London.

Frazer, J. G. 1922. The Golden Bough: A Study in Magic and Religion, abridged ed. MacMillan and Co., New York.

Frazer, J. G. 1955a. The Golden Bough: A Study in Magic and Religion, 3rd ed., Part I: The Magic Art and the Evolution of Kings, vol. 2. McMillan and Co., St. Martin's Press, London.

Frazer, J. G. 1955b. The Golden Bough: A Study in Magic and Religion, 3rd ed., Part II: Taboo and the Perils of the Soul. McMillan and Co., St. Martin's Press, London.

Frazer, J. G. 1955c. The Golden Bough: A Study in Magic and Religion, 3rd ed., Part III: The Dying God. McMillan and Co., St. Martin's Press, London.

Frazer, J. G. 1955d. The Golden Bough: A Study in Magic and Religion, 3rd ed., Part V: Spirits of the Corn and of the Wild. McMillan and Co., St. Martin's Press, London.

Frazer, J. G. 1955e. The Golden Bough: A Study in Magic and Religion, 3rd ed., Part VII: Balder the Beautiful. McMillan and Co., St. Martin's Press, London.

Freeman, P. 2006. The Philosopher and the Druids. Simon & Schuster, New York.

Frisvad, J. C., and R. A. Samson. 2004. Polyphasic taxonomy of *Penicillium* subgenus *Penicillium*: A guide to identification of food and air-borne terverticillate Penicillia and their mycotoxins. Studies in Mycology 49:1–173.

Gallagher, L. 2005. Stone Age beer. Discover 26(11); www.discover.com.

Gams, W., and J. A. Stalpers. 1994. Has the prehistoric ice-man contributed to the preservation of living fungal spores? FEMS Microbiology Letters 120:9–10.

Gandilian, P. A. 1998. Archaeobotanical evidence for evolution of cultivated wheat and barley in Armenia. Pages 280–285 in: The Origins of Agriculture and Crop Domestication: The Harlan Symposium, ed. A. B. Damania, J. Valkoun, G. Wilcox, and C. O. Qualset. Published jointly by ICARDA, IPGRI, FAO, and University of California. Report 21 of Genetic Resources Conservation Program, Division of Agriculture and Natural Resources, University of California.

Gaster, T. H. 1946. A Canaanite ritual drama: The spring festival at Ugarit. Journal of the American Oriental Society 66:49–76.

Geller, J. 1992. From prehistory to history: Beer in Egypt. Pages 19–26 in: The Followers of Horus: Studies Dedicated to Michael Allen Hoffman 1944–1990, ed. R. Friedman and B. Adams. Egyptian Studies Association Publication no. 2, Oxbow Monograph 20. Oxbow Books, Oxford.

Geml, J., G. A. Laursen, K. O'Neill, H. C. Nusbaum, and D. L. Taylor. 2006. Beringian origins and cryptic speciation events in the fly agaric (*Amanita muscaria*). Molecular Ecology 15:225–239.

Gepts, P. 2004. Crop domestication as a long-term selection experiment. Plant Breeding Reviews 24:1–44.

Gimbutas, M. 1982. The Goddesses and Gods of Old Europe: 6500–3500 BC; Myths and Images. University of California Press, Berkeley.

Gimbutas, M. 1991. The Civilization of the Goddess: The World of Old Europe. Harper Collins, New York.

Gimbutas, M. 1999. The Living Goddesses, ed. and suppl. M. R. Dexter. University of California Press, Berkeley.

Gliwa, B. 2003. Witches in Baltic fairy tales. Onomasiology Online 4:1–14.

Goldschmidt, L. 1897. Der babylonische Talmud. Calcary, Berlin.

Goor, A. 1967. The history of the pomegranate in the Holy Land. Economic Botany 21:215–229.

Gorham, L. D., and V. M. Bryant. 2001. Pollen, phytoliths, and other microscopic plant remains in underwater archaeology. International Journal of Nautical Archaeology 30(2):282–298.

Grant, M. 1970. Nero. Weidenfeld and Nicolson, London.

Grant, M. 1975. The Twelve Caesars. Simon & Schuster, New York.

Graves, R. 1958. Steps: Stories, Talks, Essays, Poems, Studies in History. Cassell & Co., London.

Graves, R. 1960. The Greek Myths, vols. 1 and 2. Penguin Books, Harmondsworth, Middlesex, U.K.

Graves, R. 1966. The White Goddess, amended and enlarged ed. Farrar, Straus and Giroux, New York.

Grendon, F. 1909. The Anglo-Saxon charms. Journal of American Folklore 22(84):105–237.

Grimm-Samuel, V. 1991. On the mushroom that deified the emperor Claudius. Classical Quarterly 41:178–182.

Grivetti, L. E. 2001. Mediterranean food patterns: The view from antiquity; ancient Greeks and Romans. Pages 3–30 in: The Mediterranean Diet: Constituents and Health Promotion, ed. A.-L. Matalas, A. Zampelas, V. Stavrinos, and I. Wolinksy. CRC Press, Boca Raton, FL.

Guasch-Jane, M. R., M. Ibern-Gómez, C. Andrés-Lacueva, O. Jáurgui, and R. M. Lamuela-Raventós. 2004. Liquid chromatography with mass spectrometry in tandem mode applied for the identification of wine markers in residues from ancient Egyptian vessels. Analytical Chemistry 76:1672–1677.

Guzmán, G., J. W. Allen, and J. Gartz. 1998. A worldwide geographical distribution of the neurotropic fungi, an analysis and discussion. Annali de Museo Civico de Rovereto Sez.: Arch., St., Sc. nat. 14:189–280.

Haldane, C. W. 1990. Shipwrecked plant remains. Biblical Archaeologist March:55–60.

Haldane, C. 1991. Recovery and analysis of plant remains from some Mediterranean shipwreck sites. Pages 213–223 in: New Light on Early Farming: Recent Developments in Paleoethnobotany, ed. J. M. Renfrew. Edinburgh University Press, Edinburgh.

Haldane, C. W. 1993. Direct evidence for organic cargoes in the late Bronze Age. World Archaeology 24(3):348–360.

Halperin, D. M. 1983. The forebearers of Daphnis. Transactions of the American Philological Association 113:183–200.

Harlan, J. R. 1976. Diseases as a factor in plant evolution. Annual Review of Phytopathology 14:31–51.

Harrison, J. E. 1903. Prolegomena to the Study of Greek Religion. Cambridge University Press, Cambridge.

Harrison, J. E. 1962. Epilegomena to the Study of Greek Religion; and Themis: A Study of the Social Origins of Greek Religion. University Books, New Hyde Park, NY (reprinting in a combined volume of Epilegomena, Cambridge, 1921, and Themis, 2nd ed., Cambridge, 1912).

Haselwandter, K., and M. R. Ebner. 1994. Microorganisms surviving for 5300 years. FEMS Microbiology Letters 116:189–193.

Hawkins, L. K. 1999. Microfungi associated with the banner-tailed kangaroo rat, *Dipodomys spectabilis*. Mycologia 91:735–741.

Hawks, C. 2001. Historical survey of the sources of contamination of ethnographic materials in museum collections. Collection Forum 16:2–11.

Hayden, B. J. 2003. The final Neolithic-early Minoan I/IIA settlement history of the Vrokastro area, Mirabello, eastern Crete. Mediterranean Archaeology and Archaeometry 3:31–44.

Hayes, J. W. 1909. Deneholes and other chalk excavations: Their origin and uses. Journal of the Royal Anthropological Institute of Great Britain and Ireland 39:44–76.

Heather, P. 2006. The Fall of the Roman Empire: A New History of Rome and the Barbarians. Oxford University Press, New York.

Heinrich, C. 2002. Magic Mushrooms in Religion and Alchemy. Park Street Press, Rochester, VT.

Helbaek, H. 1958. Grauballemandens sidste mältid. Kuml 1958:83–116 (Århus).

Heller, R. M., T. W. Heller, and J. M. Sasson. 2003. Mold: *"Tsara'at,"* Leviticus, and the history of a confusion. Perspectives in Biology and Medicine 46:588–591.

Herm, G. 1975. The Celts: The People Who Came out of Darkness. St. Martin's Press, New York.

Hershkovitz, I., M. Speirs, A. Katznelson, and B. Arenburg. 1992. Unusual pathological condition in the lower extremities of a skeleton from ancient Israel. American Journal of Physical Anthropology 88:23–26.

Hill, R. A., J. Lacey, and P. J. Reynolds. 1983. Storage of barley grain in Iron Age type underground pits. Journal of Stored Products Research 19(4):163–172.

Hofmann, A. 1972. Ergot—a rich source of pharmaceutically active substances. Pages 236–260 in: Plants in the Development of Modern Medicine, ed. T. Swain. Harvard University Press, Cambridge, MA.

Hornsey, I. S. 2003. A History of Beer and Brewing. Royal Society of Chemistry, Cambridge.

Houghton, W. 1885. Notices of fungi in Greek and Latin authors. Annals and Magazine of Natural History 15(5° ser.):22–49.

Howard, D. H., I. Weitzman, and A. A. Padhye. 2003. Onygenales: Arthrodermataceae. Pages 141–194 in: Pathogenic Fungi in Humans and Animals, 2nd ed., ed. D. H. Howard. Marcel Dekker, New York.

Huang, H.-T. 2000. Fermentations and Food Science. Part 5 of Biology and Biological Technology, Vol. 6 of Science and Civilization in China, ed. J. Needham. Cambridge University Press, Cambridge.

Hulme, P. E. 2004. Islands, invasions and impacts: A Mediterranean perspective. Pages 337–361 in: Island Ecology, ed. J. M. Fernandez Palacios and C. Morici. Asociación Española de Ecología Terrestre, La Laguna, Spain.

Husselman, E. M. 1952. The granaries of Karanis. Transactions and Proceedings of the American Philological Association 83:56–73.

Imholtz, A. A. Jr. 1977. Fungi and place names; the origin of Boletus. American Journal of Philology 98:71–76.

Innes, J. B., and J. J. Blackford. 2003. The ecology of late Mesolithic woodland disturbances: Model testing with fungal spore assemblage data. Journal of Archaeological Science 30:185–194.

Ishida, H. 2002. Insight into ancient Egyptian beer brewing using current folkloristic methods. MBAA TQ (Master Brewers Association of the Americas) 39:81–88.

Jacobsen, T. 1976. The Treasures of Darkness: A History of Mesopotamian Religion. Yale University Press, New Haven.

James, T. G. H. 1962. The Hekanakhte Papers and Other Early Middle Kingdom Documents. Metropolitan Museum of Art, New York.

Janińska, B. 2002. Historic buildings and mould fungi: Not only vaults are menacing with "Tutankhamen's curse." Foundations of Civil and Environmental Engineering 2:43–54.

Järv, R. n.d. "The three suitors of the king's daughter": Character roles in the Estonian versions of the dragon slayer (AT300). www.folklore.ee/folklore/vol22/dragons.pdf.

Jay, M. 2004. Psychedelica Victoriana. Fortean Times, February 2004; www.forteantimes.com.

Jennings, J., K. L. Antrobus, S. J. Atencio, E. Glavich, R. Johnson, G. Loffler, and C. Luu. 2005. "Drinking beer in a blissful mood": Alcohol production, operational chains, and feasting in the ancient world. Current Anthropology 46(2):275–303.

Jensen, J. 1998. The heroes: Life and death. Pages 88–97 in: Gods and Heroes of the European Bronze Age, ed. K. Demakopoulou, C. Eluère, J. Jensen, A. Jockenhövel, and J.-P. Mohen. Thames and Hudson, London.

Jenson, I. 1998. Bread and baker's yeast. Pages 172–198 in: Microbiology of Fermented Foods, 2nd ed., ed. B. J. B. Wood. Blackie Academic & Professional, London.

Jiménez-Gasco, M. M., J. A. Navas-Cortés, and R. M. Jiménez-Díaz. 2004. The *Fusarium oxysporum* f. sp. *ciceris/Cicer arietinum* pathosystem: A case study of the evolution of plant-pathogenic fungi into races and pathotypes. International Microbiology 7:95–104.

Johnsson, L. 1990. Brandkorn i Bibeln, stinksot i vetet och *Tilletia:* litteraturen-en kortfattad historik från svensk horisont. Växtskyddsnotiser 54:76–80.

Johnston, B. 2003. Experts recreate the Pompeii wine praised by Pliny. Daily Telegraph, April 24; www.telegraph.co.uk.

Jürgenson, A. n.d.a. The shit of treasure bearer, the fallen down clouds and the heavenly mushrooms. Translated summary from Mäetagused 4, 1997; www.folklore.ee/tagused/nr4/summ.htm.

Jürgenson, A. n.d.b. On popular mycology and psychotropic fungi in industrial society. Translated excerpt from Mäetagused 27, 2004; www.folklore.ee/tagused/nr27/summ.htm.

Kane, J. 1997. A historical perspective. Pages xv-xvii in: Laboratory Handbook of Dermatophytes by J. Kane, R. Summerbell, L. Sigler, S. Krajden, and G. Land. Star Publishing Co., Belmont, CA.

Kaplan, R. W. 1975. The sacred mushroom in Scandinavia. Man 10(1):72–79.

Kavaler, L. 1965. Mushrooms, Molds and Miracles. John Day, New York.

Kearns, E. 1989. Review of Persephone's Quest: Entheogens and the Origins of Religion. The Classical Review, New Series 39:400–401.

King, J. C. 1970. A Christian View of the Mushroom Myth. Hodder & Stoughton, London.

Kislev, M. E. 1982. Stem rust of wheat 3300 years old found in Israel. Science 216:993–994.

Kislev, M. E. 1991. Archaeobotany and storage archaeoentomology. Pages 121–136 in: New Light on Early Farming: Recent Developments in Paleoethnobotany, ed. J. M. Renfrew. Edinburgh University Press, Edinburgh.

Koch, H. A. 1983. Zum Nachweis einiger mikrobiologischer Probleme im Buch Leviticus (3. Buch Mose) des Alten Testamentes. Mykosen 26:159–162.

Kochkina, G. A., N. E. Ivanushkina, S. G. Karasev, E. Y. Gavrish, L. V. Gurina, L. I. Evtushenko, E. V. Spirina, E. A. Vorob'eva, D. A. Giliichinsii, and S. M. Ozerskaya. 2001. Survival of micromycetes and actinobacteria under conditions of long-term natural cryopreservation. Microbiology (Nauka/Interperiodica) 70:356–364.

Kozlovskii, A. G., V. P. Zhelifonova, V. M. Adanin, T. V. Antipova, S. M. Ozerskaya, G. A. Kochkina, and U. Gräfe. 2003. The fungus *Penicillium citrinum* Thom 1910 VKM FW-800 isolated from ancient permafrost sediments as a producer of the ergot alkaloids agroclavine-1 and epoxyagroclavine-1. Microbiology (MAIK Nauka Interperiodica) 72:723–727.

Kramer, S. N. 1956. From the Tablets of Sumer. Falcon's Wing Press, Indian Hills, CO.

Kramer, S. N. 1963. The Sumerians. University of Chicago Press, Chicago.

Kwon-Chung, K., and J. E. Bennett. 1992. Medical Mycology. Lea & Febiger, Philadelphia.

Lacey, J. 1972. The microbiology of grain stored underground in Iron Age type pits. Journal of Stored Products Research 8:151–154.

Laing, L. 1979. Celtic Britain. Charles Schribner's Sons, New York.

Lamari, L., S. E. Strelkov, A. Yahyaoui, J. Orabi, and R. B. Smith. 2003. The identification of two new races of *Pyrenophora tritici-repentis* from the host center of diversity confirms a one-to-one relationship in tan spot of wheat. Phytopathology 93:391–396.

Lamberg-Karlovsky, C. C. 2002. Archaeology and language: The Indo-Iranians. Current Anthropology 43:63–88.

Landsberger, B., and T. Jacobsen.1955. An old Babylonian charm against Mehru. Journal of Near Eastern Studies 14:14–21.

Lapatin, K. D. S. 2002. Mysteries of the Snake Goddess. Houghton Mifflin, Boston.

Large, E. C. 1940. The Advance of the Fungi. H. Holt and Co., New York.

Latacz, J. 2004. Troy and Homer: Towards a Solution of an Old Mystery. Oxford University Press, Oxford.

Lenné, J. M., and D. Wood. 1991. Plant diseases and the use of wild germplasm. Annual Review of Phytopathology 29:35–63.

Leonard, K. J., and L. J. Szabo. 2005. Stem rust of small grains and grasses caused by *Puccinia graminis*. Molecular Plant Pathology 6(2):99–111.

Leppik, E. E. 1970. Gene centers of plants as sources of disease resistance. Annual Review of Phytopathology 8:323–344.

Lévi-Strauss, C. 1970. Les champignons dans la culture. L'homme 10:5–16.

Lewis, G. 1987. A lesson from Leviticus: Leprosy. Man (New Series) 22:593–612.

Lewis-Williams, D., and D. Pearce. 2005. Inside the Neolithic Mind. Thames and Hudson, London.

Leyland, C. G. 2003 (reprint of 1897). Etruscan Magic and Occult Remedies. Kessinger, Whitefish, MT.

Lindberg, D. C. 1992. The Beginnings of Western Science: The European Scientific Tradition in Philosophical, Religious, and Institutional Context, 600 B.C. to A.D. 1450. University of Chicago Press, Chicago.

Lindow, J. 2000. Berserks. Pages 39–40 in: Medieval Folklore: A Guide to Myths, Legends, Tales, Beliefs, and Customs, ed. C. Lindahl, J. McNamara, and J. Lindow. Oxford University Press, Oxford.

Littleton, C. S. 1986. The Pneuma Enthusiastikon: On the possibility of hallucinogenic "vapors" at Delphi and Dodona. Ethos 14:76–91.

Lovatelli, E. C. 1879. Un vaso cinerario de marmo con rappresentanze relative ai Misteri de Eleusi. Bollettino della Commissione Archaeologica Comunale de Roma, pp. 3–16.

Lowenstein, E. J. 2004. Paleodermatoses: Lessons learned from mummies. Journal of the American Academy of Dermatology 50:919–936.

Luck, G. 2001. Review of The Road to Eleusis. American Journal of Philology 122:135–138.

Lydolph, M. C., J. Jacobsen, P. Arctander, M. T. P. Gilbert, D. A. Gilichinsky, A. J. Hansen, E. Willerslev, and L. Lange. 2005. Beringian paleoecology inferred from permafrost-preserved fungal DNA. Applied and Environmental Microbiology 71:1012–1017.

Lyons, D., and A. C. D'Andrea. 2003. Griddles, ovens, and agricultural origins: An enthnoarchaeological study of bread making in highland Ethiopia. American Anthropologist 105(3):515–530.

Ma, L.-J., S. O. Rogers, C. M. Catranis, and W. T. Starmer. 2000. Detection and characterization of ancient fungi entrapped in glacial ice. Mycologia 92:286–295.

Maenchen-Helfen, O. J. 1973. The World of the Huns: Studies in Their History and Culture. University of California Press, Berkeley.

Marangudakis, M. 2004. Harmony and tension in early human ecology: From prosopocentrism to early theocentrism. Human Ecology Reviews 11:133–152.

Mark, S. 2002. Alexander the Great, seafaring, and the spread of leprosy. Journal of the History of Medicine 57:285–311.

Marketos, S. G., and C. N. Ballas. 1982. Historical perspectives: Bronchial asthma in the medical literature of Greek antiquity. Journal of Asthma 19:263–269.

Marler, J. 2001. The legacy of Marija Gimbutas: An archaeomythological investigation of the roots of European civilization. Pages 89–115 in: Le Radici Prime dell' Europa: Gli Intrecce Genetici, Linguistici, Storici, ed. G. Bocchi and M. Ceruti. Bruno Mondadori, Milan.

Marmion, V. J., and T. E. J. Wiedemann. 2002. The death of Claudius. Journal of the Royal Society of Medicine 95:260–261.

Marota, I., and F. Rollo. 2002. Molecular paleontology. Cellular and Molecular Life Sciences 59:97–111.

Marr, J. S., and C. D. Malloy. 1996. An epidemiological analysis of the ten plagues of Egypt. Caduceus 12:7–24.

Marshall, A. 1999. Magnetic prospection at high resolution: Survey of large silo-pits in Iron Age enclosures. Archaeological Prospection 6:11–29.

Marshall, D., B. Tunali, and L. R. Nelson. 1999. Occurrence of fungal endophytes in species of wild *Triticum*. Crop Science 39:1507–1512.

Marthari, M. 1998. Cycladic marble idols: The silent witnesses of an island society in the early Bronze Age Aegean. Pages 159–163 in: Gods and Heroes of the European Bronze Age, ed. K. Demakopoulou, C. Eluère, J. Jensen, A. Jockenhövel, and J.-P. Mohen. Thames and Hudson, London.

Matossian, M. K. 1989. Poisons of the Past: Molds, Epidemics and History. Yale University Press, New Haven.

Matthews, R. 1997. Correspondence: Sacred circle. Nature 388:822.

Mattingly, H. 1967. Tacitus on Britain and Germany: A Translation of the "Agricola" and the "Germania." Penguin Books, Baltimore.

Maude, R. B. 1996. Seedborne Diseases and Their Control: Principles and Practice. CAB International, Wallingford, Oxon., U.K.

McCann, A. M. 2001. An early imperial shipwreck in the deep sea off Skerki Bank. Rei Cretariae Romanae Favtorvm Acta 37:257–264.

McGee, H. 1984. On Food and Cooking: The Science and Lore of the Kitchen. Charles Scribner's Sons, New York.

McGovern, P. E. 2003. Ancient Wine: The Search for the Origins of Viniculture. Princeton University Press, Princeton.

McGovern, P. E., J. Zhang, J. Tang, Z. Zhang, G. R. Hall, R. A. Moreau, A. Nuñez, E. D. Butrym, M. P. Richards, C.-S. Wang, G. Cheng, Z. Zhao, and C. Wang. 2004. Fermented beverages of pre- and proto-historic China. Proceedings of the National Academy of Sciences 101:17593–17598.

McKenna, T. 1988. Hallucinogenic mushrooms and evolution. ReVision 10(4):51–57.

McKenna, T. 1992. Food of the Gods: The Search for the Original Tree of Knowledge; A Radical History of Plants, Drugs, and Human Evolution. Bantam, New York.

McLaughlin, D. J., E. G. McLaughlin, and P. A. Lemke (eds.). 2001a. The Mycota, Vol. VII: Systematics and Evolution, Part A. K. Esser and P. A. Lemke, ser. eds. Springer, Berlin.

McLaughlin, D. J., E. G. McLaughlin, and P. A. Lemke (eds.). 2001b. The Mycota, Vol. VII: Systematics and Evolution, Part B. K. Esser and P. A. Lemke, ser. eds. Springer, Berlin.

Merlin, M. D. 2003. Archaeological evidence for the tradition of psychoactive plant use in the Old World. Economic Botany 57(3):295–323.

Merlini, M. 2006. The Gradešnica script revisited. Acta Terrae Septimcastrensis 5:25–78.

Meskell, L. 2000. Masquerades and mis/representations: Or when is a triangle just a triangle? Cambridge Archaeological Journal 10:370–372.

Michelot, D., and L. M. Melendez-Howell. 2003. *Amanita muscaria:* Chemistry, biology, toxicology, and ethnomycology. Mycological Research 107:131–146.

Mills, J. O. 1992. Beyond nutrition: Antibiotics produced through grain storage practices, their recognition and implications for the Egyptian Predynastic. Pages 27–35 in: The Followers of Horus: Studies Dedicated to Michael Allen Hoffman 1944–1990, ed. R. Friedman and B. Adams. Egyptian Studies Association Publication no. 2, Oxbow Monograph 20. Oxbow Books, Oxford.

Mitchell, S. 2004. Gilgamesh: A New English Version. Free Press, New York.

Moe, D., and O. Rackham. 1992. Pollarding and a possible explanation of the Neolithic elmfall. Vegetation History and Archaeobotany 1:63–68.

Moldenke, H. N., and A. L. Moldenke. 1952. Plants of the Bible. The Ronald Press, New York.

Molitoris, H. P. 2002. Pilze in Medizin, Folklore und Religion. Feddes Repertorium 113:165–182.

Money, N. P. 2002. Mr. Bloomfield's Orchard: The Mysterious World of Mushrooms, Molds, and Mycologists. Oxford University Press, New York.

Money, N. P. 2007. The Triumph of the Fungi: A Rotten History. Oxford University Press, New York.

Monot, M., N. Honoré, T. Garnier, R. Araoz, J.-Y. Coppée, C. Lacroix, S. Sow, J. S. Spencer, R. W. Truman, D. L. Williams, R. Gelber, M. Virmond, B. Flageul, S.-N. Cho, B. Ji, A. Paniz-Mondolfi, J. Convit, S. Young, P. E. Fine, V. Rasolofo, P. J. Brennan, and S. L. Cole. 2005. On the origin of leprosy. Science 308:1040–1042.

Monthoux, O., and K. Lundström-Baudais. 1979. Polyporaceae from the Neolithic sites of Clairvaux and Charavines France. Candollea 34:153–168.

Mooney, J. 1887. The medical mythology of Ireland. Proceedings of the American Philosophical Society 24:136–166.

Morgan, A. 1995. Toads and Toadstools: The Natural History, Folklore, and Cultural Oddities of a Strange Association. Celestial Arts, Berkeley.

Mortimer, R. K. 2000. Evolution and variation of the yeast (*Saccharomyces*) genome. Genome Research 10:403–409.

Morton, A. G. 1981. History of Botanical Science. Academic Press, London.

Moule, L. T. 1990. "History of Veterinary Medicine" from Bulletin Societe Centrale de Medicine Veterinaire (Recueil de' Medicine Veterinaire) 1890–1923, trans. C. Olson, R. Kastelic, J. Kastelic, and J. Ostroff. Typescript.

Murray, M. A. 1955. Ancient and modern ritual dances in the Near East. Folklore 66:401–409.

Museo d'Arte e Scienza. n.d. Determining the authenticity of excavated objects of art: A section of the Museo d'Arte e Scienza (Milan); www.excavatedartauthenticity.com.

Musselman, L. J. 2000. *Zawan* and tares in the Bible. Economic Botany 54:537–542.

Nakassis, D., and K. Pluta. 2003. Linear A and multidimensional scaling. Pages 335–342 in Metron: Measuring the Aegean Bronze Age, Aegaeum 24, ed. K. P. Foster and R. Laffineur. University of Texas, Austin.

Nash, M. J. 1985. Crop Conservation and Storage in Cool Temperate Climates. Pergamon, Oxford.

Nees von Esenbeck, C. G. 1816–1817. Das System der Pilze und Schwämme. Ein Versuch. Mit 44 nach der Natur ausgemalten Kupfertafeln, und einigen Tabellen. Stahelschen Buchhandlung, Würzburg.

Nilsson, M. P. 1949. A History of Greek Religion. Clarendon Press, Oxford.

North, J. A. 2003. Liber Pater. In: The Oxford Classical Dictionary, ed. S. Hornblower and A. Spaworth. Oxford University Press, New York.

Novick, R. M., and D. J. Brown. 1992 (n.d. online). Mushrooms, elves and magic: Interview with Terrence McKenna. In Mavericks of the Mind, Internet ed., www.levity.com/mavericks/terrence.htm.

Nutton, V. 2004. Ancient Medicine. Taylor & Francis, London.

Onnela, J., T. Lempiäinen, and J. Luoto. 1996. Viking age cereal cultivation in SW Finland—A study of charred grain from Pahamäki in Pahki, Lieto. Annales Botanici Fennici 33:237–255.

Orlob, G. B. 1971. History of plant pathology in the Middle Ages. Annual Review of Plant Pathology 9:7–20.

Orlob, G. B. 1973. Ancient and medieval plant pathology. Pflanzenschutz-Nachrichten 26:61–281.

Otto, W. F. 1965. Dionysus: Myth and Cult. Indiana State University Press, Bloomington.

Pääbo, S., H. Poinar, D. Serre, V. Jaenicke-Després, J. Hebler, N. Rohland, M. Kuch, J. Krause, L. Vigilant, and M. Hofreiter. 2004. Genetic analyses from ancient DNA. Annual Review of Genetics 38:645–679.

Pahlow, G., R. E. Muck, F. Driehuis, S. J. W. H. Oude Elferink, and S. K. Spoelstra. 2003. Microbiology of ensiling. Pages 31–93 in: Silage Science and Technology, ed. D. R. Buxton, R. E. Muck, and J. H. Harrison. Agronomy Monographs no. 42. American Society of Agronomy, Crop Science Society of America, and Soil Science Society of America, Madison, WI.

Palaima, T. G. 2004a. Sacrificial feasting in the Linear B documents. Hesperia 73:217–246.

Palaima, T. G. 2004b. Appendix One: Linear B sources. Pages 439–454 in: Anthology of Classical Myth: Primary Sources in Translation, ed. S. M. Trzaskoma, R. S. Smith, and S. Brunet. Hackett, Indianapolis.

Panagiotakopulu, E., and P. C. Buckland. 1991. Insect pests of stored products from late Bronze Age Santorini, Greece. Journal of Stored Products Research 27:179–184.

Parker, A. J. 1992. Ancient Shipwrecks of the Mediterranean and the Roman Provinces. British Archaeological Reports Int. Ser. 580. Tempus Reparatum, Oxford.

Paterson, J. 1982. "Salvation from the sea": Amphorae and trade in the Roman west. Journal of Roman Studies 72:146–157.

Patrich, J. n.d. Warehouses and granaries in Caesaria Maritima. http://research .haifa.ac.il/~archlgy/patrichj/warehouse/warehouse.html.

Pegler, D. N. 2000a. Useful fungi of the world: Stone fungus and fungus stones. Mycologist 14(3):98–101.

Pegler, D. N. 2000b. Useful fungi of the world: Agaricum—The "universal remedy" of ancient Rome. Mycologist 14(4):146–147.

Pegler, D. N. 2001. Useful fungi of the world: Amadou and chaga. Mycologist 15(4):153–154.

Pegler, D. N. 2002. Useful fungi of the world: The "poor man's truffles of Arabia" and "manna of the Israelites." Mycologist 16(1):8–9.

Peintner, U., R. Pöder, and T. Pümpel. 1998. The iceman's fungi. Mycological Research 102:1153–1162.

Perles, C. 2001. The Early Neolithic in Greece. Cambridge University Press, Cambridge.

Perry, I., and P. D. Moore. 1987. Dutch elm disease as an analogue of Neolithic elm decline. Nature 326:72–73.

Petrakis, V. 2004. Ship representations on late Helladic III C pictorial pottery: Some notes. Inferno 9:1–6.

Pharand, M. 2005. Poetic mythography: The genesis, rationale and reception of The Greek Myths. Gravesiana 8; gravesiana.robertgraves.org/content/ view/8, retrieved April 30, 2005.

Phillips, J. 1982. Fungi and ancient medicine. Histoire des Sciences Medicales 17(Spec. 2):204–208.

Pike, J. H. 1938. Frivolities of Courtiers and Footprints of Philosophers: Being a Translation of the First, Second, and Third Books and Selections from the Seventh and Eighth Books of the Policraticus of John of Salisbury. Reprint, 1972. Octagon Books, New York.

Pinhasi, R., and M. Pluciennik. 2004. A regional biological approach to the spread of farming in Europe: Anatolia, the Levant, south-eastern Europe, and the Mediterranean. Current Anthropology 45(Supp.):S59–S82.

Pini, R. 2004. Late Neolithic vegetation history at the pile-dwelling site of Palu de Livenza (northeastern Italy). Journal of Quaternary Science 19:769–781.

Piperno, D. R., E. Weiss, I. Host, and D. Nadel. 2004. Processing of wild cereal grains in the Upper Paleolithic revealed by starch grain analysis. Nature 430:670–673.

Plantzos, D. 1997. Crystals and lenses in the Graeco-Roman world. American Journal of Anthropology 101:451–464.

Pliny. 1945. Natural History, IV, trans. H. Rackham. Loeb Classical Library, Harvard University Press, Cambridge, MA.

Pollington, S. 2003. The Mead-Hall: The Feasting Tradition in Anglo-Saxon England. Anglo-Saxon Books, Norfolk, U.K.

Pollock, S. 2003. Feasts, funerals, and fast food in early Mesopotamian states. Pages 17–38 in: The Archaeology and Politics of Food and Feasting in Early States and Empires, ed. T. L. Bray. Kluwer Academic/Plenum, New York.

Postgate, J. N. 1992. Early Mesopotamia: A Society and Economy at the Dawn of History. Routledge, London.

Powell, J. U. 1929. Rodent-gods in ancient and modern times. Folklore 40:173–179.

Purcell, N. 1985. Wine and wealth in ancient Italy. Journal of Roman Studies 75:1–19.

Purdy, L. H. 1968. The microflora of Tushka tufa. Pages 934–935 in: The Prehistory of Nubia, ed. F. Wendorf. Fort Burgwin Research Center, Southern Methodist University Press, Dallas.

Ramoutsaki, I. A., C. E. Papadakis, I. A. Ramoutsakis, and E. S. Helidonis. 2002. Therapeutic methods used for otolaryngological problems during the Byzantine period. Annals of Otology, Rhinology and Laryngology 111:553–557.

Ramsbottom, J. 1953. Mushrooms and Toadstools. Frontis, Collins, London.

Reiner, E. 1995. Astral magic in Babylonia. Transactions of the American Philosophical Society, New Series 85(4):1–150 + xiii.

Renfrew, C. 2000. At the edge of knowability: Towards a prehistory of languages. Cambridge Archaeological Journal 10:7–34.

Reynolds, P. J. 1976. Farming in the Iron Age. Cambridge University Press, London.

Ribes, J. A., C. L. Vanover-Sams, and D. J. Baker. 2000. Zygomycetes in human disease. Clinical Microbiology Reviews 13:236–301.

Richards, M. P. 2002. A brief review of the archaeological evidence for Paleolithic and Neolithic subsistence. European Journal of Clinical Nutrition 56:9 pp. (not paginated, www.nature.com/ejcn).

Richards, M. 2003. The Neolithic invasion of Europe. Annual Review of Anthropology 32:135–162.

Rickman, G. 1971. Roman Granaries and Store Buildings. Cambridge University Press, Cambridge.

Rickman, G. 1980. The Corn Supply of Ancient Rome. Oxford University Press (Clarendon Press), New York.

Riedlinger, T. J. 1993. Wasson's alternative candidates for soma. Journal of Psychoactive Drugs 25:149–156.

Robbins, M. 2001. Collapse of the Bronze Age: The Story of Greece, Troy, Israel, Egypt, and the Peoples of the Sea. Authors Choice, Lincoln, NB.

Robinson, M. A. 2000. Coleopteran evidence for the elm decline, Neolithic activity in woodland clearance and the use of the landscape. Pages 27–36 in: Plants in Neolithic Britain and Beyond, Neolithic Studies Group Seminar Papers 5, ed. A. S. Fairbairn. Oxbow Books, Oxford.

Rogers, S. O., and J. D. Castello. 2001. Life in ancient ice: A workshop sponsored by the National Science Foundation, 30 June–3 July 2001, Glendon

Beach, Oregon. http://salegus-scar.montana.edu/docs/workshopdocs/ AncientIc%20Report.pdf.

Rolfe, R. T., and F. W. Rolfe. 1925. The Romance of the Fungus World. Chapman & Hall. (Reprint 1974, Dover, New York.)

Rollo, F., and I. Marota. 1999. How microbial ancient DNA, found in association with human remains, can be interpreted. Philosophical Transactions of the Royal Society of London B 354:111–119.

Rollo, F., W. Asci, S. Antonini, I. Marota, and M. Ubaldi. 1994. Molecular ecology of a Neolithic meadow: The DNA of the grass remains from the archaeological site of the Tyrolean Iceman. Experientia 50:576–584.

Rollo, F., S. Sassaroli, and M. Ubaldi. 1995. Molecular phylogeny of the fungi from the Iceman's clothing. Current Genetics 28:289–297.

Rollo, F. S., M. Ubaldi, L. Ermini, and I. Marota. 2002. Ötzi's last meals: DNA analysis of the intestinal content of the Neolithic glacier mummy from the Alps. Proceedings of the National Academy of Sciences 99:12594–12599.

Rose, H. J. 1922. Lua Mater: Fire, rust, and war in early Roman cult. Classical Review 36:15–18.

Rosenthal, T. 1961. Aulus Cornelius Celsus. Archives of Dermatology 84:613–618.

Ross, W. D., ed. 1913. The Work of Aristotle: De Plantis. University Press, Oxford.

Rostovtzeff, M. 1941. The Social and Economic History of the Hellenistic World. Clarendon Press, Oxford.

Rostovtzeff, M. 1957. The Social and Economic History of the Roman Empire. Clarendon Press, Oxford.

Roussel, R., S. Rapior, C.-L. Masson, and P. Boutié. 2002. *Fomes fomentarius* (L. : Fr.) Fr. : Un champignon aux multiples usages. Cryptogamie, Mycologie 23:349–366.

Rubchak, B. 1981a. Notes on the text. Pages 43–75 in: Shadows of Forgotten Ancestors (M. Kotsiubynsky), Ukrainian Classics in Translation no. 4, ed. G. S. N. Luckyj. Published for The Canadian Institute of Ukrainian Studies by Ukrainian Academic Press, Littleton, CO.

Rubchak, B. 1981b. The music of Satan and the bedeviled world: An essay on Mykhailo Kotsiubynsky. Pages 77–121 in: Shadows of Forgotten Ancestors, Ukrainian Classics in Translation no. 4, ed. G. S. N. Luckyj. Published for The Canadian Institute of Ukrainian Studies by Ukrainian Academic Press, Littleton, CO.

Ruck, C. A. P. 1983. The offerings of the Hyperboreans. Journal of Ethnopharmacology 8:177–207.

Ruck, C. A. P. 2006. Sacred Mushroom of the Goddess: Secrets of Eleusis. Ronin Publishing, Berkeley.

Ruck, C. A. P., and D. Staples. 1994. The World of Classical Myth: Gods and Goddesses; Heroines and Heroes. Carolina Academic Press, Durham, NC.

Ruck, C. A. P., B. D. Staples, and C. Heinrich. 2001. The Apples of Apollo: Pagan and Christian Mysteries of the Eucharist. Carolina Academic Press, Durham, NC.

Rudgley, R. 1995. The archaic use of hallucinogens in Europe: An archaeology of altered states. Addiction 90:163–164.

Russell, D. 1998. Shamanism and the Drug Propaganda: The Birth of Patriarchy and the Drug War. Kalyx, Camden, NY.

Saggs, H. W. F. 2005 (reprint of 1988). The Babylonians: A Survey of the Ancient Civilization of the Tigris-Euphrates Valley. The Folio Society, London.

Şahoğlu, V. 2005. The Anatolian trade network and the Izmir region during the early Bronze Age. Oxford Journal of Archaeology 24:339–361.

Salomonsson, A. 1990. Food for thought—Themes in recent Swedish ethnological food research. Ethnologia Scandinavica 20:111–133.

Salway, P., and W. Dell. 1955. Plague at Athens. Greece & Rome, Ser. 2 2:62–70.

Samorini, G. 1992. The oldest representations of hallucinogenic mushrooms in the world (Sahara Desert, 9000–7000 B.P.). Integration 2/3:69–78.

Samorini, G., and G. Camilla. 1994. Rappresentazioni fungine nell'arte greca. Annali del Museo Civico di Rovereto 10:307–326; www.samorini.net/doc/sam/greca.html.

Sampson, A. P., and S. T. Holgate. 1997. Allergenic asthma: An interaction between inflammation and repair. Pages 7–28 in: Toxicology of Chemical Respiratory Hypersensitivity, ed. I. Kimber and R. J. Dearman. Taylor & Francis, London.

Sasson, J. M. 1994. The blood of grapes: Viticulture and intoxication in the Hebrew Bible. Pages 399–419 in: Drinking in Ancient Societies: History and Culture of Drinks in the Ancient Near East, ed. L. Milano. Sargon, Padua.

Saunders, J. B. De C. M. 1965. Review: The Medical Background of Anglo-Saxon England: A Study in History, Psychology and Folklore. Isis 56:93–95.

Scheer, F. M. 2004. Ninkasi—An ancient brew revisited. MBAA TQ 41:120–123.

Scherrer, B., E. Isidore, P. Klein, J.-S. Kim, A. Bellec, B. Chalhoub, B. Keller, and C. Feuillet. 2005. Large intraspecific haplotype variability at the *Rph7* locus results from rapid and recent divergence in the barley genome. Plant Cell 17:361–374.

Schoental, R. 1980. A corner of history: Moses and mycotoxins. Preventive Medicine 9:159–161.

Schoental, R. 1984. Mycotoxins and the Bible. Perspectives in Biology and Medicine 28:117–120.

Schoental, R. 1987. Mycotoxins, the flood, and human lifespan in the Bible. Koroth 9:503–506.

Schoental, R. 1991. Mycotoxins, porphyrias and the decline of the Etruscans. Journal of Applied Toxicology 11:453–454.

Schoental, R. 1994. Mycotoxins in food and the plague in Athens. Journal of Nutritional Medicine 4:83–85.

Schukking, S. 1976. De geschiedenis van het inkuilen. Stikstof 19:313–321.

Schultes, R. E. 1998. Antiquity of the use of New World hallucinogens. Heffter Review of Psychedelic Research 1:1–7.

Schumann, G. 1991. Plant Diseases: Their Biological and Social Impact. American Phytopathological Society, St. Paul, MN.

Schwartz, G. M., and H. H. Curvers. 1992. Tell al-Raqā'i 1989 and 1990: Further investigations at a small rural site of early urban northern Mesopotamia. American Journal of Archaeology 96:397–419.

Schwartz, M. 2002. Early evidence of reed boats from southeast Anatolia. Antiquity 76:617–618.

Scudamore, K. A., and C. T. Livesey. 1998. Occurrence and significance of mycotoxins in forage crops and silage: A review. Journal of the Science of Food and Agriculture 77:1–17.

Sharnoff, S. D. n.d. Lichens and people: Bibliographic database of the human uses of lichens. www.lichen.com/people.html.

Sinha, R. N. 1995. The stored grain ecosystem. Pages 1–32 in: Stored Grain Ecosystems, ed. D. S. Jayas, N. D. G. White, and W. E. Muir. Marcel Dekker, New York.

Sledzik, P. A. 2001. Nasty little things: Molds, fungi, and spores. Pages 71–77 in: Dangerous Places: Health, Safety, and Archaeology, ed. D. A. Poirier and K. L. Feder. Bergin & Garvey, Westport, CT.

Smith, A. E., and D. M. Secoy. 1975. Forerunners of pesticides in classical Greece and Rome. Journal of Agricultural and Food Chemistry 23:1050–1055.

Smith, C. C., and O. J. Reichman. 1984. The evolution of food caching by birds and mammals. Annual Review of Ecology and Systematics 15:329–351.

Smith, H. 1972. Wasson's "Soma"—A review article. Journal of the American Academy of Religion 40:480–499.

Smith, S. T. 2003. Pharoahs, feasts and foreigners: Cooking, foodways, and agency on ancient Egypt's southern frontier. Pages 39–64 in: The Archaeology and Politics of Food and Feasting in Early States and Empires, ed. T. L. Bray. Kluwer Academic/Plenum, New York.

Solomon, J. 1995. The Apician sauce, *ius apicianum*. Pages 115–131 in: Food in Antiquity, ed. J. Wilkins, D. Harvey, and M. Dobson. University of Exeter Press, Exeter, U.K.

Soyer, A. 1853. The Pantropheon: Or, A History of Food and Its Preparation in Ancient Times. Simpkin, Marshall, London.

Spivey, N. 1996. Etruscomania and Etruscosense. American Journal of Archaeology 100:170–173.

Sprengel, K. 1822. Theophrast's Naturgeschichte der Gewächse, Hammerich, Altona, vols. 1 and 2.

Stankiewicz, E. 1958. Slavic kinship terms and the perils of the soul. Journal of American Folklore 71(280):115–122.

Stefanaki, I., E. Foufa, A. Tsatsou-Dritsa, and P. Dais. 2003. Ochratoxin A concentrations in Greek domestic wines and dried vine fruits. Food Additives and Contaminants 20:74–83.

Stewart, R. B., and W. Robertson. 1968. Fungus spores from prehistoric potsherds. Mycologia 60:701–704.

Sturtevant, A. M. 1952. Etymological comments upon certain Old Norse proper names in the Eddas. PMLA 67:1145–1162.

Suhr, E. G. 1974. Winged words. Folklore 85:169–172.

Taylor, T. N., and J. M. Osborn. 1996. The importance of fungi in shaping the paleoecosystem. Review of Paleobotany and Palynology 90:249–262.

Temin, P. 2003. Mediterranean trade in Biblical times. MIT Department of Economics Working Paper Series, Working Paper 03–12. 24 pp.

Theophrastus. 1916a. Inquiry into Plants, and Minor Works on Odours and Weather Signs, I, trans. A. Hort. Loeb Classical Library, G. P. Putnam's Sons, New York.

Theophrastus. 1916b. Inquiry into Plants, and Minor Works on Odours and Weather Signs, II, trans. A. Hort. Loeb Classical Library, G. P. Putnam's Sons, New York.

Theophrastus. 1990a. De Causis Plantarum II, Books III-IV, ed. and trans. B. Einarson and G. K. K. Link. Loeb Classical Library, Harvard University Press, Cambridge, MA.

Theophrastus. 1990b. De Causis Plantarum III, Books V-VI, ed. and trans. B. Einarson and G. K. K. Link. Loeb Classical Library, Harvard University Press, Cambridge, MA.

Thiselton-Dyer, T. F. 1994 (reprint of 1889). The Folklore of Plants. Llanerch Publishers, Lampeter, U.K.

Thomas, C. G., and C. Conant. 2005. The Trojan War. Greenwood Press, Westport, CT.

Thompson, S. 1955–1958. Motif-Index of Folk-Literature. Vols. 1–6. Indiana University Press, Bloomington.

Thompson, S. 1977. The Folktale. University of California Press, Berkeley.

Toporov, V. N. 1985. On the semiotics of mythological conceptions about mushrooms. Trans. S. Rudy. Semiotica 53–4:295–357.

Toussaint-Samat, M. 1992. History of Food, trans. A. Bell. Blackwell, Cambridge, MA.

Trease, G. E., and W. C. Evans. 1978. Pharmacognosy, 11th ed. Baillière Tindall, London.

Tufnell, B. O. 1924. Czecho-Slovak folklore. Folklore 35:26–56.

Turner, R. C., M. Rhodes, and J. P. Wild. 1991. The Roman body found on Grewelthorpe Moor in 1850: A reappraisal. Britannia 22:191–201.

Twede, D. 2002. The packaging technology and science of ancient transport amphoras. Packaging Technology and Science 15:181–195.

Urtāns, J. 1992. Cylindrical ritualistic stones with a cup mark in Latvia. Journal of Baltic Studies 23:47–56.

Uther, H.-J. 2004. The Types of International Folktales: A Classification and Bibliography Based on the System of Antii Aarne and Stith Thompson. Suomalainen Tiedeakatemia, Academia Scientarum Fennica, Helsinki.

Vaidya, J. G., and A. S. Rabba. 1993. Fungi in folk medicine. Mycologist 7:131–133.

Valamoti, S. M., and P. C. Buckland. 1995. An early find of *Oryzaephilus surinamensis* (L.) (Coleoptera: Silvanidae) from final Neolithic Mandalo, Macedonia, Greece. Journal of Stored Products Research 31:307–309.

van Geel, B., J. Buurman, O. Brinkkemper, J. Schelvis, A. Aptroot, G. van Reenan, and T. Hakbijl. 2003. Environmental reconstruction of a Roman Period settlement site in Uitgeest (The Netherlands), with special reference to coprophilous fungi. Journal of Archaeological Science 30:873–883.

Varro, Marcus Terentius. 1934. On Agriculture, trans. W. E. Hooper. Loeb Classical Library, Harvard University Press, Cambridge, MA.

Verster, A. J. 1976. Ergotism in ancient Greece. American Journal of Physical Anthropology 44(1):213.

Vilenskaya, L. n.d. From Slavic mysteries to contemporary PSI research and back, Part 3: Where myth merges with reality: Slavic mysteries. http://resonateview.org/places/writings/larissa/myth.htm.

Vinokurova, N. G., N. E. Ivanushkina, G. A. Kochkina, M. U. Arinbasarov, and S. M. Ozerskaya. 2005. Production of mycophenolic acid by fungi of the genus *Penicillium* Link. Applied Biochemistry and Microbiology 41:83–86.

Virgil. 2002. Georgics, trans. K. Chew. Hackett Publishing Company, Indianapolis.

Vogt, E. Z. 1958. Review: Mushrooms, Russia and History. American Antiquity 24:85–86.

Wainwright, M. 1989a. Moulds in ancient and more recent medicine. The Mycologist 3:21–23.

Wainwright, M. 1989b. Moulds in folk medicine. Folklore 100:162–166.

Ward, M. P., and L. L. Couëtil. 2005. Climatic and aeroallergen risk factors for chronic obstructive pulmonary disease in horses. American Journal of Veterinary Research 66:818–824.

Warner, E. A. 2002. Death by lightning: For sinner or saint? Beliefs from Novosokol'niki Region, Pskov Province, Russia. Folklore 113:248–259.

Wasson, R. G. 1968. Soma: Divine Mushroom of Immortality. Harcourt Brace Jovanovich, New York.

Wasson, R. G. 1971. The soma of the Rig Veda: What was it? Journal of the American Oriental Society 91:169–187.

Wasson, R. G. 1972. What was the soma of the Aryans? Pages 201–213 in: Flesh of the Gods, ed. P. T. Furst. Praeger, New York.

Wasson, R. G. 1980. The Wondrous Mushroom: Mycolatry in Mesoamerica. McGraw-Hill, New York.

Wasson, R. G., A. Hofmann, and C. A. P. Ruck. 1978. The Road to Eleusis: Unveiling the Secret of the Mysteries. Harcourt Brace Jovanovich, New York.

Wasson, R. G., S. Kramrisch, J. Ott, and C. A. P. Ruck. 1986. Persephone's Quest: Entheogens and the Origin of Religion. Yale University Press, New Haven.

Wasson, V. P., and R. G. Wasson. 1957. Mushrooms, Russia and History, vols. 1 and 2. Pantheon, New York.

Webster, P., D. M. Perrine, and C. A. P. Ruck. 2000. Mixing the Kyleon. Eleusis: Journal of Psychoactive Plants and Compounds, New Series 4:1–25.

Wicklow, D. T. 1995. The mycology of stored grain: An ecological perspective. Pages 197–249 in: Stored Grain Ecosystems, ed. D. S. Jayas, N. D. G. White, and W. E. Muir. Marcel Dekker, New York.

Wiencke, M. H. 1989. Change in early Helladic II. American Journal of Archaeology 93:495–509.

Wijbenga, A., and O. Hutzinger. 1984. Chemicals, man and the environment: A historic perspective of pollution and related topics. Naturwissenschaften 71:239–246.

Wilde, W. R. 1979 (reprint of 1852). Irish Popular Superstitions. Irish Academic Press, Dublin.

Wilkinson, J. M., K. K. Bolsen, and C. J. Lin. 2003. History of silage. Pages 1–30 in: Silage Science and Technology, ed. D. R. Buxton, R. E. Muck, and J. H. Harrison. Agronomy Monographs no. 42. American Society of Agronomy, Crop Science Society of America, and Soil Science Society of America, Madison, WI.

Wilkinson, T. J. 1994. The structure and dynamics of dry-farming states in Upper Mesopotamia. Current Anthropology 35:483–520.

Willerslev, E., and A. Cooper. 2005. Ancient DNA. Proceedings of the Royal Society of London B 272:3–16.

Willerslev, E., A. J. Hansen, B. Christensen, J. P. Steffensen, and P. Arctander. 1999. Diversity of Holocene life forms in fossil glacial ice. Proceedings of the National Academy of Sciences 96:8017–8021.

Williams, N. 1996. How the ancient Egyptians brewed beer. Science 273:432.

Wilson, D. G. 1975. Plant remains from the Graveney Boat and the early history of *Humulus lupulus* L. in W. Europe. New Phytologist 75:627–648.

Wilson, J. V. K. 1994. The samanu disease in Babylonian medicine. Journal of Near Eastern Studies 53:111–115.

Wilson, P. L. 1999. Ploughing the Clouds: The Search for Irish Soma. City Lights, San Francisco.

Wood, E. 1989. World Sourdoughs from Antiquity. Sinclair Publishing, Cascade, ID.

Wood, J. 2000. Food and drink in European prehistory. European Journal of Archaeology 3:89–111.

Wood, M. 1985. In Search of the Trojan War. Facts on File, New York.

Wood-Martin, W. G. 1970 (reprint of 1902). Traces of the Elder Faiths of Ireland: A Folklore Sketch; a Handbook of Irish Pre-Christian Traditions, vols. 1 and 2. Kennikat Press, Port Washington, NY.

Worobec, C. D. 1995. Witchcraft beliefs and practices in pre-revolutionary Russian and Ukrainian villages. Russian Review 54:165–187.

Wyand, R. A., and J. K. M. Brown. 2003. Genetic and forma specialis diversity in *Blumeria graminis* of cereals and its implications for host-pathogen co-evolution. Molecular Plant Pathology 4:187–198.

Yeats, W. B. 1973. Fairy and Folk Tales of Ireland. Reprint of Fairy and Folk Tales of the Irish Peasantry plus Irish Fairy Tales, both first published in 1892. Macmillan, New York.

Yiannikouris, A., and J.-P. Jouany. 2002. Mycotoxins in feeds and their fate in animals: A review. Animal Research 51:81–99.

Zaffarano, P. L., B. A. McDonald, M. Zala, and C. C. Linde. 2006. Global hierarchical gene diversity analysis suggests the Fertile Crescent is not the center of origin of the barley scald pathogen *Rhynchosporium secalis*. Phytopathology 96:941–950.

Zapata, L., L. Peña-Chocarro, G. Pérez-Jordá, and H.-P. Stika. 2004. Early Neolithic agriculture in the Iberian Peninsula. Journal of World Prehistory 18:283–325.

Zhan, J., R. E. Pettway, and B. A. McDonald. 2003. The global genetic structure of the wheat pathogen *Mycosphaerella graminicola* is characterized by high nuclear diversity, low mitochondrial diversity, regular recombination, and gene flow. Fungal Genetics and Biology 38:286–297.

Zohary, D., and M. Hopf. 2000. Domestication of Plants in the Old World, 3rd ed. Oxford University Press, Oxford.

Zohary, M. 1982. Plants of the Bible. Cambridge University Press, Cambridge.

Index

137